T0143118

Progress in IS

"PROGRESS in IS" encompasses the various areas of Information Systems in theory and practice, presenting cutting-edge advances in the field. It is aimed especially at researchers, doctoral students, and advanced practitioners. The series features both research monographs that make substantial contributions to our state of knowledge and handbooks and other edited volumes, in which a team of experts is organized by one or more leading authorities to write individual chapters on various aspects of the topic. "PROGRESS in IS" is edited by a global team of leading IS experts. The editorial board expressly welcomes new members to this group. Individual volumes in this series are supported by a minimum of two members of the editorial board, and a code of conduct mandatory for all members of the board ensures the quality and cutting-edge nature of the titles published under this series.

More information about this series at http://www.springer.com/series/10440

Rimvydas Skyrius

Business Intelligence

A Comprehensive Approach to Information
Needs, Technologies and Culture

 Springer

Rimvydas Skyrius
Economic Informatics
Vilnius University
Vilnius, Lithuania

ISSN 2196-8705   ISSN 2196-8713   (electronic)
Progress in IS
ISBN 978-3-030-67034-4   ISBN 978-3-030-67032-0   (eBook)
https://doi.org/10.1007/978-3-030-67032-0

This Springer imprint is published by the registered company Springer Nature Switzerland AG.
The registered company address is: Gewerbestrasse 11, 6330 Cham, Switzerland

Foreword

Business has always been an early adopter of advanced management techniques and technologies, expected to provide significant support to value creation. Over several decades of the use of digital information technologies in business, however, success stories have been mixed with disappointments and confusing experience. This can be especially true for the subfield of business intelligence (BI) activities, where the expectations and stakes are considerably higher. Most executives and organizations are well familiar with information technologies, systems, and frameworks for their use and expected value, but quite often they are uncertain about how to best deploy them to create this value. This book by professor Rimvydas Skyrius of Vilnius University in Lithuania, *Business Intelligence: A Comprehensive Approach to Information Needs, Technologies and Culture,* summarizes much of what the author has researched and learned in the field over the years, and presents this experience to the potential audience of business executives and managers, BI professionals, and concerned researchers and students who want to take a less technology-oriented approach to the field.

In the field of business intelligence, fast and tumultuous developments bring possibilities, expectations, and a good deal of confusion to accompany them. This book comes at a time when the situation in the field experiences both significant developments and challenges. The developments are rather visible: large surge of interest in analytics, fast development of artificial intelligence techniques, and utilization of global information resources, to name a few. The challenges, however, are no less visible: despite huge technical advances, many obstacles still remain, numerous contradictions remain unsolved, and many BI projects fail or do not bring the promised benefits. As numerous research works and real-life cases show, mere possession of advanced technology and voluminous information resources is no longer considered adequate for fulfilling business information needs. This explains the rising interest in managerial and human issues.

The focus of the book is under-researched managerial and human issues in BI. The book addresses several important research gaps—relation between BI maturity and agility, role and features of BI culture, and definition of and relations between soft assets—attention, sense, and trust. By adopting a user needs-based aspect in research issues, the book also addresses several important practical needs— improved match between business and BI activities, insights of higher quality, improved BI project success rate, and clarity regarding principal sources of BI value. In addressing the aforementioned issues, the book attempts at reduction of confusion around these issues. This focus point defines the selected set of topics and its consistency and coverage. The proposed set of topics attempts to cover the managerial and human issues that are essential in sense making and insight development: positioning of BI in overall informing activities, possible BI dimensions, information integration issues, management of experience and lessons learned, relation of BI maturity and agility, features of BI culture, and role of soft BI assets. The author steps carefully in a rather vague and ill-defined field of managerial and human issues of BI by relating phenomena to their origins and history, and deliberately leaves technology issues in the background. The author also uses his own former research, published in numerous previous research papers and conferences, to give ground to presented arguments.

It may be possible to argue with the author on some statements presented in the book. For example, the author declares that there exists a distinct BI culture (Chapter 8), when many sources state that organizational culture and information culture are significant enough to determine whether the organizational climate is catalytic for effective informing. On the other hand, if the content of this book initiates a round of discussions on current business informing issues, the book will for sure have reached its goal.

Although the author does not attempt to present ready-made rules or recipes for advanced informing and insight development, I am confident that this book will be useful to assorted segments of the reader audience: business executives and managers, researchers working on intelligence, analytics, and other contemporary informing issues, and teachers and students from the academic community, taking interest in related problems.

Rector, University of Economics in   Celina M. Olszak, Ph.D.
Katowice, Katowice, Poland

Acknowledgements

This book reflects my beliefs about the informing activities in business that have formed drop by drop over the last 30 years, starting with decision support, later moving on to business intelligence, business analytics, and other advanced informing activities that produce insights for a business user. The important source of guiding cues for all this time has been the collected empirics targeted at business information needs—that is, the research interest has been driven much more by issues of user pull than by issues of technology push. Another source of guidance, no less important than the empirics, has been the people who talked to me, listened to my sometimes incongruous and naïve ideas, read my work, and provided advice. Although I am not able to mention every one of these wonderful people that provided shoulder in one way or another, I would like to use this occasion to mention at least those who come to mind in the first place.

First and foremost, I want to express my deepest thanks to the people at Springer who directly dealt with my book—Dr. Christian Rauscher for believing in my work, Ms. Sujatha Chakkala and Ms. Rajeswari Sathyamurthy for their patience, meticulous attention to detail, and for maintaining the overall positive background to our communication, and all the other people at the publisher whom I never met, but whose solid professionalism was felt all the time.

I also want to use this occasion and thank my numerous colleagues whose support has been vital over many years. First of all, these are my direct colleagues from the Department of economic informatics at the Vilnius University, and many ideas have been first tested among them. My year-long exchange stay with the University of Illinois at Chicago has been quite a while ago, but the people over there helped a lot to get our sometimes messy body of knowledge into order and provided guidelines that stand until now. Special thanks here go to Professors Sharon Reeves, Aditya Saharia, Aris Ouksel, Yair Babbad, Bronius Vaškelis, and late Irena Baleisis. Heartfelt thanks go to Dr. Eli Cohen and Elizabeth Boyd of Informing Science Institute, together with Professors Terence Grandon Gill and Michael Jones, and Professors Tom Wilson and Elena Maceviciute of Information Research for supporting my modest findings. I also wish to thank my PhD students and all other students for their often-fresh opinion on emerging issues in research. Special thanks go to Chris Butler for the courage to read my first chapters and perceive them from the business angle.

Last but not least, I want to thank my family, and in the first place my wife Ruta, for their support and patience. The time I have spent on preparing this book could have been given to family, but I had them covering my back all the time; it is largely their support in making this book come to reality.

Yours sincerely

Rimvydas Skyrius

Contents

Chapter 1
Business Intelligence: Human Issues

1.1 Introduction

The stormy and kaleidoscopic development of the field of business intelligence (BI), mostly attributed to advances in information technology (IT), has created significant confusion in the area of business informing and an imbalance between IT and human issues, to the favor of the former. Early in the era of decision support systems, Feldman and March (1981) have noted a controversy between information engineering, represented by information systems (IS), and information behavior, represented by intelligence: "some strange human behavior may contain a coding of intelligence that is not adequately reflected in engineering models". It is quite ironic that in the field of business intelligence the dominating emphasis has been made on information technology as being intelligent, while the intelligent need to be aware arises from intelligence activities of the people, and this aspect has received a lot less emphasis. The current book is a modest attempt by the author to reduce this imbalance.

No rational activity is performed without having information on its environment, and business activities are no exception. Many sources have stated that information is an important asset, together with capital, people, competencies, know-how and the like. However, such statements emerged almost exclusively with the growth and spread of IT, while information already has been important for centuries. The specific nature of business information lies in the ability to organize and coordinate all other assets. This makes information a very special resource overarching other resources, and therefore possessing special importance. To organize this information efficiently, information systems are developed that use contemporary information technology and advanced processing logic to handle information properly. Any information system is a unity of people and technology; essentially, information systems are systems of people and relations between people, and technology role comes after that. Business intelligence is one of such systems that aims at supporting deeper and more complete understanding of activity and its environment by

R. Skyrius, *Business Intelligence*, Progress in IS,
https://doi.org/10.1007/978-3-030-67032-0_1

employing tools and methods of advanced informing. The need for this kind of systems has been recognized by Nobel prize winner Herbert Simon as early as 1971: "Designing an intelligence system means deciding: when to gather information (much of it will be preserved indefinitely in the environment if we do not want to harvest it now); where and in what form to store it; how to rework and condense it; how to index and give access to it; and when and on whose initiative to communicate it to others" (Simon, 1971).

The objective of the book. As stated above, the current book is an attempt to balance the human and technology aspects of BI to their fair shares. "Fair", of course, is a subjective concept in the eye of the beholder, but a certain experience, spent by the author researching business decision support and BI, has formed a belief that many troubles haunting BI installations originate from human and managerial issues. It would be unfair to blame BI technologies, developed by highly creative and professional people in the software development field, for BI failures. Digital technology does exactly what it is intended to do, and failure to create expected value is most likely a result of inflated expectations and inadequate preparations. Meanwhile, according to Gill (2015), large white spots remain in understanding how people and organizations perform intelligence. The orientation towards the virtues of technology has prevailed for a long time and continues to do so, while the direction towards the understanding of the client and community that are actual users of the information has been receiving considerably less attention.

There are many other areas where intelligence activities are the axis of the job, and BI obviously has its roots in informing activities from other fields—scientific research, political and military intelligence, law enforcement etc. BI is not a new concept, and deemed by some to be a fading one. Yet, when assessed against its potential to impact business awareness, coordination and insightfulness, it is not likely to fade anytime, as the same can be said about any intelligence activities. From the emergence of the term, the field has been rather confusing and torn apart by confronting opinions and discussions on terms and boundaries. As a marketing term, BI may have lost some of its flair, but the need to perform intelligence activities is not going to disappear. So if this book moves the attention of researchers and practitioners just a little to the space of human issues in BI, it's goal will be achieved.

1.2 Novelty and Originality

The book axis joins together several key aspects:

- BI is an activity whose prime feature is the boost of human intelligence potential by supporting it appropriately with contemporary IT;
- BI is a test-bed for sophisticated IT and avant-garde business applications;
- A prolonged phase of accentuated technology features as the main driver for BI advancement has introduced an imbalance of attention;
- Human and managerial issues deserve a shift in research approaches;

- Advanced informing is firstly about human aspects, and secondly—about effective use of IT.

The field of advanced informing has experienced its own series of twists and turns. After the jump from **knowledge** issues (knowledge management wave) to **data** issues (Big Data wave with support from artificial intelligence and machine learning), probably the time has come for researchers and practitioners to assign due attention to **information** issues. Users expect from BI to become well-informed, the system delivering right information to the right person at the right time (Cohen, 1999; Fischer, 2012; Hayes-Roth, 2007). The principal product of BI is high-value information and insights to cope with important, costly and often non-routine problems. There are numerous issues complicating this task:

- Growing complexity of information environment, recently amplified by the Big Data phenomena and its increasing 3Vs—volume, velocity, variety;
- Limited power of routine information systems (ERP and similar systems) to handle this complexity and the entire spectrum of information needs;
- Contradiction between information needs and reluctance to use available systems, mostly due to additional time and attention required (failures of knowledge management, lessons-learned and similar systems);
- Goals and preferences of participants/stakeholders are diverse, conflicting and most often are understood only during decision process;
- BI effort is often directed at production of information, not at supporting its effective use;
- Problems of integration of information from different sources;
- Importance of soft information and its resistance to processing.

According to Boyer, Frank, Green, Harris, and Van De Vanter (2010), top-performing organizations will establish both a vision about the strategic use of information and a plan to implement this vision. To develop this vision, BI has to balance efficiency and flexibility, just like any other information system. In the case of BI, efficiency is expected in storing, accessing and integrating data; providing access to data and analytical tools that are simple and easy to use; presenting results in clear form. Flexibility is required to cope with the multitude of information sources, formats, analytical approaches and procedures, communication and delivery channels to properly serve the variety of possible problems to come.

The possible elements of novelty of approach in this book are:

- A detailed analysis of business information needs;
- Relations between business dimensions and BI dimensions;
- Balance of hard (IT) and soft (human) factors;
- Analysis of and relations between BI maturity and agility;
- Definition of position and features of BI culture;
- Discussion of important soft BI factors like attention, sense and trust.

The structure of the book has been largely dictated by the research logic:

- The first three chapters are dedicated to the positioning of BI: Chap. 2 presents definitions and clarifications on what is considered BI in this book; Chap. 3 discusses relations of BI with other kinds of IS and informing activities; and Chap. 4 projects the variety of business dimensions into BI implementation dimensions;
- The next two chapters deal with information integration: Chap. 5 discusses issues of data and information integration and relations between the two, while Chap. 6 is dedicated to the issues of lessons learned and experience preservation;
- Technology variety and features are briefly presented in Chap. 7—as BI technology is not the key point in this book, the available technologies are given a brief overview;
- The issues of BI maturity and agility, being important in understanding the dynamics of BI implementation, adoption and use, are discussed in Chap. 8;
- The issues of BI culture are discussed in Chap. 9, while the soft factors of attention, sense and trust are discussed in Chap. 10;
- Lastly, the notion of holistic or encompassing BI, as well as BI education and research issues are discussed in Chap. 11.

1.3   The Target Audience and Benefits

The target audience for this book may be quite wide in its variety, from undergraduate students to researchers and practitioners working in the field. This is, of course, a rather vague description of target population, but a more focused definition can be provided. Firstly, there is a community of researchers that work in related directions of BI implementation, acceptance and value, BI maturity and agility, BI culture and informing culture in general. Being grateful for the opportunity to get familiar with their work, the author hopes that this book might give something in return. Secondly, there is an obvious growth of study programs at all levels from undergraduate to doctoral studies, related in one or other way to BI and business analytics, and this book may improve the balance of technical and managerial subjects. Thirdly, under the current wave of interest in all aspects of analytics—Big Data, artificial intelligence, machine learning and the like—this book might provide a few points when formulating analytical questions or making sense of results.

References

Boyer, J., Frank, B., Green, B., Harris, T., & Van De Vanter, K. (2010). *Business intelligence strategy: A practical guide for achieving BI excellence*. Ketchum, ID: MC Press Online.
Cohen, E. B. (1999). Reconceptualizing information systems as a field of the discipline informing science: From ugly duckling to swan. *Journal of Computing and Information Technology, 7*(3), 2013–2219.

Feldman, M. S., & March, J. G. (1981). Information in organizations as signal and symbol. *Administrative Science Quarterly, 26*(2), 171–186.

Fischer, G. (2012). Context-aware systems: The 'right' information, at the 'right" time, in the 'right' place, in the 'right' way, to the 'right' person. *AVI'12: Proceedings of the International Working Conference on Advanced Visual Interfaces*, May 21–25, 2012, Capri Island, Italy, pp. 287–294.

Gill, T. G. (2015). *Informing science* (Vol. 1). Santa Rosa, CA: Informing Science Press.

Hayes-Roth, F. (2007). *Valued information at the right time (VIRT): Why less volume is more value in hastily formed networks*. Retrieved March 13, 2020, from https://calhoun.nps.edu/bitstream/handle/10945/36760/Hayes-Roth_VIRT_why_less_volume_2006.pdf?sequence=1&isAllowed=y.

Simon, H. A. (1971). Designing organizations for an information-rich world. In M. Greenberger (Ed.), *Computers, communication, and the public interest* (pp. 37–72). Baltimore, MD: The Johns Hopkins Press.

Chapter 2
Business Intelligence Definition and Problem Space

2.1   The Activities of Business Intelligence

The current business information environment is becoming more and more compli-cated for many reasons, detailed analysis of which goes well beyond the scope of this book. There are multiple pressures on organizations from dynamic business envi-ronment, and the subsequent need for proper actions creates significant information challenges. A substantial variety of business environment factors is creating pres-sures on businesses, to name a few:

- Global issues—capital and labor migration, financial and commodity markets;
- Government regulation, or regulation by international governing bodies;
- Market forces—customer needs and demand, competition, market dynamics;
- Technology landscape—innovations, substitutes, disruptive technologies;
- Growing dynamism, rate of innovations, changes and needs to respond etc.

A popular BI textbook (Sharda, Delen, & Turban, 2014: 27) presents a brief description of the current business informing climate: "Organizations, private and public, are under pressures that force them to respond quickly to changing conditions and to stay innovative in the way they operate. Such activities require organizations to be agile and to make frequent and quick strategic, tactical, and operational decisions, some of which are very complex". The above description pictures infor-mation environment as some kind of a moving target, where the accentuated quickness of reaction is expected to conflict with informing quality. A tradeoff emerges between them, driven by a need to maintain informing quality under changing conditions. One of the key goals of the people responsible for the informing activities is to soften the urgency and reduce complexity by designing and maintaining a comprehensive informing environment.

The need for companies to be aware of their business environment is a concept that is well understood, appreciated, and well represented in the literature (Calof & Wright, 2008). More than 40 years ago, Porter (1980) reported that whilst companies

© Springer Nature Switzerland AG 2021
R. Skyrius, *Business Intelligence*, Progress in IS,
https://doi.org/10.1007/978-3-030-67032-0_2

were carrying this activity out informally, in his opinion this was nowhere near sufficient. He advocated the need for a structured intelligence process at all times in order to continuously and systematically identify business opportunities and threats in the complexity of environment. This complexity creates both opportunities and problems that need solutions, but these opportunities and problems have to be discovered preferably on time, or at least not too late. While the need to be informed and avoid surprises is featured for a long time in any rational activities, Porter's statement in the above quote is calling for a business intelligence policy that would support the permanent and systematic nature of BI activities. Such policy would require a better understanding of BI as a key business function.

The boundaries of BI, as understood in this book, may differ from those named in other sources that often concentrate around simple reporting and internal data. Several factors that define BI as a special informing activity are:

– Its positioning is clearly on the complex part of information needs scale intended to satisfy the needs for insights and decision support that arise from important and possibly costly issues;
– Use of sophisticated approaches and tools;
– The regular information systems do not deal with issues like intelligence community and insight sharing; these factors are hardly mentioned when discussing implementation of ERP systems;
– Need to integrate multiple information from heterogeneous internal and external sources.

Business intelligence, as a distinct kind of informing activities, is obviously past its turbulent growth as a fresh concept. However, its importance for well-informed activities and awareness is hardly questionable. BI implementation and adoption already has a substantial body of experience, and it can be expected that newly developed BI systems will further maintain the principal function of providing the users with required meaningful information at required time and in required place. Same can be said about the role of any information system, but BI usually is focused on information needs that range in their complexity levels from medium to high.

Many things can serve as nasty surprises in today's business—unexpected mergers or acquisitions, aggressive pricing by competitor, surprise innovations in competing or substitute products, changes in consumer preferences (Tyson, 2006). There are many more examples of unexpected events that can make a business vulnerable. Gilad (2004) presented results of his research among the middle-level managers of Fortune 500 companies, interviewing them about the role of surprises in their activities. Two key questions have been asked regarding anticipation of unexpected events in the near future, and the actual surprises experienced by the businesses in the last 5 years (Table 2.1).

The answers to the first question clearly show the managers' awareness of turbulent business environment and the competitive pressures. However, a whole 92% of the interviewed managers have acknowledged that their companies experienced significant surprises affecting their market position. According to Gilad, this has happened not because of ignorance, but more due to accepted informing

Table 2.1 Role of surprises in corporate activities (Gilad, 2004)

1st question: To what extent does your company anticipate an increased level of business risk in its markets and industries in the next 2–3 years?	Very likely	44.1%
	Likely	36.3%
	Somewhat likely	17.6%
	Not likely	2.0%
	I don't know	0.0%
2nd question: How many times would you say that your company was, in the past 5 years, surprised by events that had the potential for a significant impact on your long-term market position?	Never	8.0%
	1–3 times	68.0%
	More than 3 times	24.0%

practices, stereotypes and limited information exchange that added to the failures to notice the signs of approaching threats. Obviously, it is impossible to correctly interpret the cues in a complex and changing environment, but it can be pointed out that the deficiencies in organizational and information culture have clearly added to the failure to make sense out of at least some of these signs.

2.2 Definitions

Many published works on BI have attempted to put together various definitions of BI in order to define the most specific features of BI. Such attempts, among other things, indicate the complexity and multi-dimensionality of BI, and it raises an issue of whether production of a perfect definition of BI is probable at all. On the other hand, BI has evolved from a status of a buzzword describing a trendy set of advanced technologies to an established type of information activity in everyday business.

Below are several well-known definitions of BI:

- Rouibah & Ould-ali, 2002: "Business intelligence is a strategic approach for systematically targeting, tracking, communicating and transforming relevant weak signs into actionable information on which strategic decision-making is based."
- Atre, 2004: "I would define business intelligence as business success realized through rapid and easy access to actionable information through timely and accurate insight into business conditions about customers, finances and market conditions."
- Wixom & Watson, 2010: "Business intelligence is an umbrella term that is commonly used to describe the technologies, applications, and processes for gathering, storing, accessing, and analyzing data to help users make better decisions."

Although the definition and boundaries of BI is and most likely will be the subject of discussion for the foreseeable future (Arnott & Pervan, 2014; Bucher, Gericke, & Sigg, 2009; Popovic, Turk, & Jaklic, 2010; Power, 2013), we assume that BI may be defined as the organizational practice that encompasses a coherent set of people, informing processes and conventions of using a comprehensive technology platform to satisfy business information needs that range from medium to high complexity. Then the objective of BI, though the exact wording varies in different definitions, is to ensure awareness and provide valuable **insights** to business managers/decision makers.

Compared to other types of information systems, business intelligence is regarded as a much younger term, with authorship assigned to Howard Dressner of Gartner Group in 1989, or even earlier—to Hans Peter Luhn in 1958 for publishing a paper titled *A Business Intelligence System*. However, an interesting fact is that we can have a retrospective look at the mission of management information systems (MIS), whose role of keeping management aware of the state of business has never been downplayed, and earlier mission definitions for MIS sound very much like the mission definitions for business intelligence today. A few explanations of MIS role from earlier sources are presented below:

- "A management information system refers to many ways in which computers help managers to make better decisions and increase efficiency of an organization operation" (Pick, 1986).
- "For information to be useful for managerial decision making, the right information (not too much and not too little) must be available at the right time, and it must be presented in the right format to facilitate the decision at hand" (Gremillion & Pyburn, 1988).
- "A management information system is a business system that provides past, present, and projected information about a company and its environment. MIS may also use other sources of data, particularly data about the environment outside of the company itself." (Kroenke, McElroy, Shuman, & Williams, 1986).
- "The systems and procedures found in today's organizations are usually based upon a complex collection of facts, opinions and ideas concerning the organization's objectives. ... For an organization to survive, it must learn to deal with a changing environment effectively and efficiently. To accomplish the making of decisions in an uncertain environment, the firm's framework of systems and procedures must be remodeled, refined, or tailored on an ongoing basis." (Fitz-Gerald, FitzGerald, & Stallings, 1981).
- "An information system can be defined technically as a set of interrelated components that collect (or retrieve), process, store, and distribute information to support decision making and control in an organization." (Laudon & Laudon, 2006).

Many definitions of business intelligence do not differ much from the above definitions; e.g., Vuori (2006) states that "... business intelligence is considered to be a process by which an organization systematically gathers, manages, and analyzes information essential for its functions". Davenport (2005) defines business intelligence as "IT applications that help organizations make decisions using technology for reporting and data access, as well as analytical applications". Gartner IT glossary (2017), paraphrasing the definition of Wixom and Watson (2010), states that "BI is an umbrella term for the technologies, applications and processes associated with collecting, storing, using, disclosing and analyzing data to facilitate sound decision making". The overlapping definitions create confusion and mix up expectations regarding the range of basic to advanced support functions. In order to have a more precise definition of business intelligence, we have to decide whether all informing functions are „intelligence", because they increase awareness, or does BI have a clear separation from other, more simple informing functions. If so, the separation criteria between BI systems and any other management information systems have to be defined, even if all systems serve the informing needs of the same organization. The presented definitions stress decision assistance that is usually attributed to decision support systems (DSS) and environments intended to serve more complex information needs. BI apparently serves the same segment of complex information needs as DSS do; in addition to supporting decisions for separate problems, BI is expected to provide awareness, although on a more sophisticated level than MIS.

More insight on definitions is given in James Kobielus blog (Kobielus, 2010), where he takes an attempt to handle the variety of data and information management aspects that various sources relate to BI:

- Narrow versus wide definition of BI: narrow is the scope as seen by the users or consumers, and wider scope reflects all the supply (data sources) and processing (tools and techniques) aspects;
- Business intelligence versus business analytics;
- Business intelligence versus competitive, market and other kinds of intelligence;
- Analysis of structured versus unstructured information;
- Historical versus predictive perspective;
- Traditional, data-based BI versus social media as an important pool of behavioral information.

We can note that these points deal with a sample of what can be called the implementation dimensions or focus dimensions of business intelligence. More attention to such dimensions will be given in Chap. 4 of this book.

It is interesting to note that, while different positions regarding BI definition are obvious, multiple definitions can be found in the same sources—e.g., Negash and Gray (2003) state that "BI systems combine data gathering, data storage, and knowledge management with analytical tools to present complex and competitive information to planners and decision makers"; the same source says "Business intelligence is a form of knowledge". Popovič, Hackney, Coelho, and Jaklič (2012) present two angles of BI in a single paper: "technological solution offering data integration and analytical capabilities to provide stakeholders at various organizational levels with valuable information for their decision-making" and "quality information in well-designed data stores, coupled with business-friendly software tools that provide knowledge workers timely access, effective analysis and intuitive presentation of the right information, enabling them to take the right actions or make the right decisions". Atre (2004) states: "I would define business intelligence as business success realized through rapid and easy access to actionable information through timely and accurate insight into business conditions about customers, finances and market conditions"; and further in the text "BI is intelligence based on business information, past actions and options available for the future". Because of the multifaceted nature of a complex phenomenon that BI is, this multitude of definitions probably shouldn't be considered controversial.

The variety of BI features leads to several different aspects of how BI is understood:

2.2.1 As an Information System (Fig. 2.1)

Figure 2.1 presents a generic picture of BI as an information system whose principal function is to present the multitude of required intelligence results to the users. The system is intended to cover a variety of information sources, including internal sources (ERP, CRM, SCM and other systems) and external sources. ETL procedures load internal data into a data warehouse, and its content is integrated with external data if necessary. The analytics step performs a variety of analytical procedures from

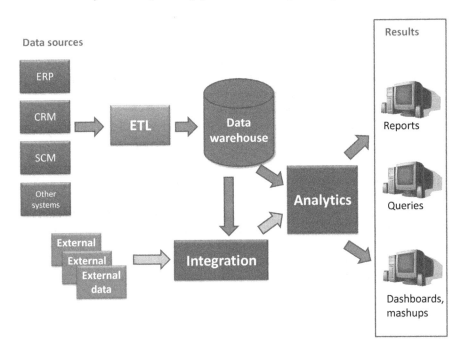

Fig. 2.1 The view of business intelligence as an information system (adapted from (Skyrius & Bujauskas, 2011))

simple to sophisticated, and produces results in the forms of reports, serviced queries, dashboards or mashups with visualized results.

2.2.2 As a Cyclical Process (Fig. 2.2)

In this case, BI process is seen as an iterative loop which covers the intelligence process in several steps from initial definition of information needs to utilization of results and feedback on possible improvements.

2.2.3 As a Technology Platform or Technology Stack (Fig. 2.3)

In Fig. 2.3, the technology platform approach places BI resources in several layers, starting with the data layer at the bottom containing the raw content, then the analytics, or "action" layer in the middle, where the analytical applications convert data into information and insights; and the presentation, "delivery" layer which serves the end users with results in presentable and easy-to-understand format.

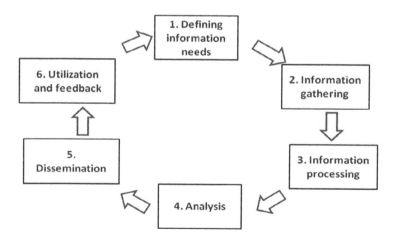

Fig. 2.2 Business intelligence as a cyclical process (Vuori, 2006)

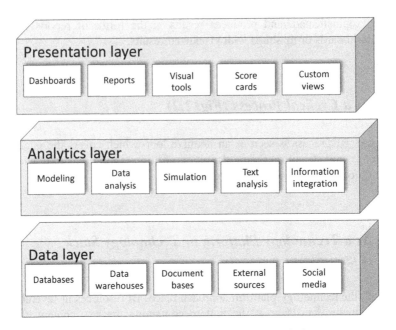

Fig. 2.3 Business intelligence as a stack of technologies (source: author)

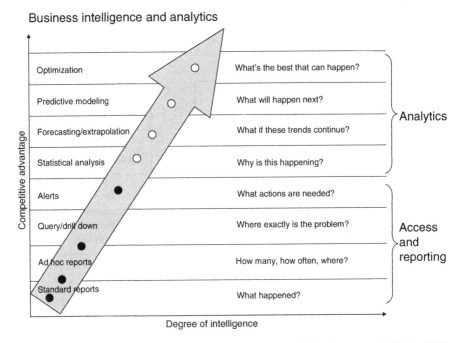

Fig. 2.4 Business intelligence as a value chain of informing activities (Davenport & Harris, 2007)

2.2.4 *As a Value Chain of Informing Activities that Covers Issues from Simple to the Most Complex (Fig. 2.4)*

Here (Fig. 2.4) BI is shown as an informing process crossing multiple levels of increasing complexity both in questions and related results. Some sources consider this process as a reflection of BI value chain that starts with simple needs and approaches, and proceeds to the most complex needs and their satisfaction.

All of the above aspects accentuate the need to reach expected BI sophistication by executing processes of increasing complexity, using advanced techniques, and intending to deliver a product of high value. The total set of such informing processes, techniques and approaches is proposed to be named *advanced informing*. This term will be discussed in more detail further in this chapter.

2.3 Expected Results and Benefits

The definitions of BI in the previous paragraph accentuate that users have substantial value expectations from using insights produced by BI. Below in Table 2.2 there are some examples of specific questions of above-than-average complexity that are

Table 2.2 Specific questions for BI environment (author, based on (Ziama & Kasher, 2004))

Analytical application	Business question	Business value
Customer base segmentation	What market segments come up in our customer base, and what are their principal features?	Customize communication with customers based on their specific features and needs
Basket analysis	What goods are purchased together? Why?	Discover relations and rules to enhance cross-promotion and cross-selling
Customer profitability	What is the lifetime profitability of a customer? What profitability dynamics are inside customer life cycle?	Discover ways of managing and boosting the overall profitability of customers
Fraud detection	How can I tell which transactions or actors are likely to be fraudulent?	Determine fraud early and take immediate action to minimize losses, or prevent fraud altogether
Customer attrition	Which customers are likely to leave?	Prevent loss of high-value customers and let go of lower-value customers
Competitor monitoring	What competitors will likely be the sources of serious problems in the near future?	Well-targeted preventive measures to protect market position and future potential
Innovation monitoring	What technology (or market) innovations are expected to bring the largest impact to the industry?	Minimized risk in decisions concerning innovations

expected to be dealt with by BI environment. Such questions require sophistication that is difficult for operations-level IS like ERP to achieve.

When defining the goals of BI, various sources present them differently, e.g.:

- Gathering and analysis of business information (Riabacke, Larsson, & Danielson, 2011);
- Timely and more accurate reporting presented to planners and decision makers (Negash & Gray, 2003; Top Business Intelligence Trends, 2017);
- A comprehensive view of activities and environment (Ramakrishnan, Jones, & Sidorova, 2012; Sabherwal & Fernandez, 2011).
- Collaboration and insight development (Atre, 2004; Boyer, Frank, Green, Harris, & Van De Vanter, 2010; Evelson, 2010; Meredith, Remington, O'Donnell, & Sharma, 2012);
- Decision-making (Arnott & Pervan, 2014; Daly, 2016; Evelson, 2010; Imhoff & White, 2008; Kandogan, Balakrishnan, Haber, & Pierce, 2014; Meredith et al., 2012; Ramakrishnan et al., 2012; Rouibah & Ould-ali, 2002; Watson & Wixom, 2007);
- Profitability and building shareholder value (Atre, 2004; Dawson & Van Belle, 2013).

The presented definitions differ in their direction and focus, but are united by their drive to provide comprehensive informing to the users, and the term *advanced informing*, used earlier in this chapter, seems quite appropriate here. We can note that the above list of functions supported by BI reflects sequential phases that are to be implemented and performed when dealing with BI solution:

- Accumulation and integration of information resources;
- Better awareness of activities and environment;
- Decisions and actions;
- Value creation.

This sequential approach is supported in (Watson & Wixom, 2007) when pointing to the sources of BI benefits: first, data integration removes redundancy and reinforces necessary common standards; then, efficient data delivery saves time at the operations level; and gradually questions evolve to important analytical and forecasting issues whose benefits are more global and difficult to quantify. Popovic et al. (2010) note that the initial stages of implementing BI usually include data organizing initiatives that reduce data inconsistency and increase data quality. For all kinds of information systems, improved data quality means higher trust in the system and a stronger enabling base for the variety of services.

One important phase that is often omitted from the sources discussing BI goals and benefits is the feedback phase that creates additional value by performing the ever-useful assessment of BI activities. Such assessment is a key to BI becoming adaptive, agile and resilient—all important features of BI value in a long run.

The expected benefits may be arranged according to specific business priorities. Boyer et al. (2010, p. 44) have suggested to categorize basic priorities in a business alignment strategy, and to work out a timeline prioritizing "must do now" vs. "invest for the future" vs. "de-prioritize." Factors influencing this prioritization include: key corporate priorities, high revenue opportunities, quick-win opportunities and others. Such approach is potentially helpful in managing the BI project priorities.

2.3.1 Difficulty in Evaluating BI Benefits

The evaluation of BI benefits pertains to a much larger research problem of estimation of benefits from investments into information technologies and systems in general. This problem, although being recognized and discussed for several decades, is still largely under-researched. The approaches to this problem are partial and incomplete, and concentrate more on naming benefit factors than measuring them.

For BI, this problem is much more complicated because of the nature of expected benefits that places them mostly in the high-stakes part of the IT-supported value spectrum.

According to a study by CDW, a technology provider (2013), there are a number of ways to extract value from BI; e.g.:

– Extract and leverage insights in near-real time;
– Eliminate guesswork;
– Offer insights into customer behavior;
– Identify cross-selling and up-selling opportunities;
– Better management of inventory;
– Gauge true manufacturing costs;
– Streamline operations.

However, we have to note that the majority of the above sources of value are quite often associated with regular information systems and regular informing that covers everyday operations, while BI, although having touch points with named sources, is expected to deliver important information of strategic or near-strategic value.

Popovic et al. (2010) discuss the estimation of BI benefits and their measurement, and state that the measurement of BI benefits does not have any widely known and tested methods, or the companies do not have resources to undertake the benefit measurement activities. Such activities, to author's opinion, would largely be pointless because of the very reasons indicated in many other sources—the near-impossibility to apply reliable measures to BI benefits that are mostly soft in their nature. A presented list of estimated empirical benefits from earlier research by the same researchers includes:

- **Rationalization of information resources**:

  – Data from different data sources is easier integrated and unified.
  – Better data quality in the transactional system as a result of identified critical points within processes.

- **Automation of simple analytic operations**:

  - Information specialists are less burdened by information preparation due to self-service access to data and now mainly maintain and expand the system, and focus more on strategic tasks.
  - Information users save time previously devoted to data preparation and analysis.
  - Unburden of analytical users, thus allowing them to focus on more complex analyses.

- **Direct support to analytical functions**:

  - Planning process on lower hierarchical levels is supported.
  - Using flexible analysis and visualization, BI allows the identification of problems that would otherwise likely be unnoticed.
  - Introduction of BIS partially contributed to a more flattened organizational structure.

Lönnqvist and Pirttimäki (2006) have attempted to generalize the approaches to measuring the value of BI and named a number of reasons why the **exact estimates of BI benefits are very difficult** or impossible altogether. The most prominent reasons, to their opinion, are the intangible nature of benefits, time lag between delivery of BI product and financial gain, difficulty in estimating the contribution of BI use to the overall benefits, subjectivity of "soft" measures like consumer satisfaction and other non-financial issues.

Watson and Wixom (2007) have pointed out that operational benefits like elimination of data redundancy are easy to measure but have local impact, while **strategic benefits** like support for achieving strategic objectives are expected to have global impact, yet **are rather difficult to quantify**.

Elbashir, Collier, and Davern (2008) have performed empirical research on recognized benefits of BI in surveyed companies and suggested estimating the business value of BI systems at two levels: business process performance level that reflects internal efficiency; and organizational performance that uses organization-wide metrics (return on investment, sales growth) to reflect competitive advantage and overall strength; the latter level is much harder to handle. Due to the difficulties in measuring the benefits, the suggested approach **uses perception-based measures** for insights into intangible benefits.

Discussing BI investments and possible benefits, Negash and Gray (2003) have pointed out that estimating expected return on BI investment is a difficult problem because **most BI benefits are soft**, and are expected to come in a form of some "big bang" return in the future. They also indicate that it would be a rare case for BI system to cover its investment through cost reductions.

The possible bottom line for the issue of BI benefit estimation is that the more ambitious and global intelligence issues arise, the more complicated is the estimation of BI benefits. One of the possible alternative approaches is to look not at the expected benefits, but at the avoidance of losses that would have been incurred without using information provided by BI. Also, in some sources it is suggested that the estimation of benefits would be shifted to the responsibility of business users who probably would more clearly see the role of BI as an enabler in business processes and decision making. A side-effect is the streamlining of the information resources, which, while not directly influencing the quality of strategic decisions, introduces a surge in information reliability that is another important enabling factor for BI value.

2.4 Advanced Informing and Insights

2.4.1 Definition of Advanced Informing, Its Positioning and Boundaries

The sources dedicated to BI issues often refer to a timeline covering the historical sequence of developments in the field of DSS, BI and other types of systems intended to serve complex information needs. According to Clark, Jones, and Armstrong (2007), the field tends to continually chase the buzz words and system types of the day, often at the expense of establishing something of greater value that transcends system type and that provides a stronger foundation for the field. Systems such as business intelligence, knowledge management, decision support, and executive information systems all support decision making and other sophisticated information needs (Arnott & Pervan, 2005), so clearly there is a thread of continuity and some common features that drive user expectations over the past several decades. These advanced features and approaches are believed to lead to what is usually defined as good informing—right information to the right person at the right time. For cases that have proved successful, it is the long-term advantageous features that drive expectations—required coverage and aggregation, reliability, data quality, rational balance of features and coverage, etc.

For better understanding and positioning of BI it may be helpful to introduce the term of advanced informing. "Advanced" is usually separated from "regular" by going for deeper, more ambitious goals, usually associated with added levels of complexity, and handling this complexity. Users expect that BI use of advanced, state-of-the-art technology will deliver advanced products satisfying complex information needs and non-routine informing.

On the other hand, complexity does not necessarily equal advanced approaches: advanced usually assumes enabling environments that provide possibilities and opportunities not associated with simpler, blunt approaches. Following this logic, complexity should be hidden from the user to create ease of use and better usability, yet complexity should be testable to develop trust in the tools of informing environment. This issue raises requirements for flexibility and usability of BI environment, directly related to self-service, acceptance and adoption. Flexibility of information environment is one of the more important things in satisfying complex information needs, as shown by earlier research (Skyrius, 2004, 2008). Coverage is another important thing—the scope of informing should encompass information sources that are internal and external, structured and unstructured, reliable and questionable, permanent and random etc.

An example of *advanced informing* at work is a convenient and easy-to-use dashboard showing important information. Although the presented information itself might look quite simple and clear, the process of producing this information would most likely be complex, encompassing a set of information sources and processing procedures.

The definition of advanced informing would be expected to relate closely to the definition of analytics. Analytics use advanced approaches and techniques that lead to (or are used for) advanced informing:

- Analytics from Excel to Big Data and deep learning;
- Information and sense integration;
- Intelligent monitoring with alerts, early warning systems, complex event processing;
- Modeling and simulation;
- Forecasts and predictions;
- Rule development and management;
- And many others.

Apart from analytical functions, advanced informing should include activities like monitoring, information gathering, insight integration, and other activities that are not exactly analytics. For the purposes of this book, we can consider analytics being a part of BI and advanced informing in general.

Advanced informing is expected to be encompassing—not only sophisticated approaches and techniques are used, but it utilizes simple functions where required as well (e.g., input from accounting systems and other base-level systems). These simple and not-so-simple functions complement the entire insight under development, so that important issues, be they simple or complex, should be assembled together.

2.4.2 Insights

According to Lönnqvist and Pirttimäki (2006), one of the principal reasons organizations implement BI is to gain better insight into its business processes, strategies, and operations. The use of the notion of insight to articulate the direction of BI efforts is supported by Kandogan et al. (2014), who state that "BI refers to the tools and processes that transform data into insights that support decision making". Companies rely on BI to make sense in a wide spectrum of needs, starting with the transactional data collected by ERP and other data intensive applications (Holsapple & Sena, 2005) and ending with complex insights to provide decision makers with a better understanding of underlying trends and dependencies that affect the business (Lönnqvist & Pirttimäki, 2006). Such understanding is expected to enable more informed and rational decision-making (Chou, Tripuramallu, & Chou,

2005). Marchand and Peppard (2013) state that the quest of heavy investment into BI is "to extract insights from the massive amounts of data now available from internal and external sources". From the above quotes in can be inferred that insight is a main product of BI, addressing important issues and therefore not easy to define, largely due to its 'soft' nature.

Vriens and Verhulst (2008) state that a business insight is: "a thought, fact, combination of facts, data and/or analysis of data that induces and furthers understanding of a situation or issue that has the potential of benefitting the business or redirecting the thinking about that situation or issue which then in turn has the potential of benefitting the business". Or, as Wills (2013) puts it, "Insight is information that can make a difference". Hence, insight is a piece of information that is highly aggregated around certain central issue and allows closing a significant gap in understanding the issue.

Dalton (2016) provides a few working definitions of an insight:

- An unrecognized fundamental human truth;
- A new way of viewing the world that causes us to reexamine existing conventions and challenge the status quo;
- A penetrating observation about human behavior that results in a fresh perspective;
- A discovery about the underlying motivations that drive people's actions.

Thomas Davenport (2014), while discussing what he calls "stories to tell with data" and what may be roughly matching the notion of insights presented in this book, distinguishes several dimensions of these stories:

1. **Time dimension**—stories may be about the past, present, or future. **Past** stories are of reporting type, summarizing the facts; **present** stories represent a current state of things and are of explanatory type; and **future** stories are predictions that use assumptions and specific methods, mostly statistics.
2. **Focus dimension**—stories may focus on the questions related to different stages of problem solving: the **what** stories report what has happened; the **why** stories look for the reasons or factors that have caused the outcome, and the **what to do** stories explore and evaluate directions to solve the problem properly. A complete story may contain all three instances of this dimension.
3. **Depth dimension**—quick stories addressing a **simple** problem; and detailed stories reflecting search for a solution to a **complex** problem.
4. **Methods dimension**—**correlation** stories describe how relationships between variables are dependent, or **causation** stories that define cause-effect relationships.

This list can easily be expanded, as it presents the features that belong not only to a problem, but to its probable solutions as well. Stories and insights, as any complex object, have multidimensionality that is essential for proper understanding of a problem under scrutiny. Such dimensions can be linked to dimensions that reflect certain alternatives of BI implementation and are discussed in one of the further chapters of this book.

A few examples of strategic business insights are:

- A group of innovators realize that they have created a potentially huge niche product;
- A marketing professional notices a unique customer need that is not satisfied by current products or services;
- An auditor notices a pattern of transactions that is "too good to be true" and points to an almost certain fraud.

A certain example of an insight is presented by McCrary (2011):

> At Greyhound, we were exploring ways to grow ridership and revenues in the Northeast corridor—between Boston, New York and Washington, DC. In our efforts to better understand what the consumer needs were that we might be able to address more effectively, we did a thorough review of our operations, looking at schedules, ridership by schedule, competition, etc. A couple of key insights were gleaned from the review: (1) Most trips between these cities were not reserved in advance—people came to the terminal and got on the next available bus; and (2) our schedules were designed to optimize our operations, not meet consumer needs. Based on these insights, we modified our schedules to operate hourly on the hour or half hour, depending on the city of origin, creating schedules that consumer could understand and count on. The change resulted in a 25% increase in revenue with no additional marketing spending.

Although seen as important truth, insight is not a momentary issue that emerges suddenly. Thomas Davenport (2015) defines insight management as a process with the phases of framing, creation, marketing, consumption, storage and reuse.

- **Framing**: what problem is to be solved? Extent of potential support by data and analytics (roughly conforms to Simon's dimension of structured/unstructured)? Do the examined alternatives cover the perceived decision space (this might change over time)? Do the analyst and decision-maker share the vision of problem solution? A possible relation between cue, context and insight might be defined—cue attracts attention, especially if it is unconventional and "out-of-place"; requires additional context to eventually develop information into insight.
- **Creation**: the problem is seen as a moving target; irrational information behaviors (Feldman & March, 1981) and side products are often useful;
- **Marketing and consumption**: based on understanding of what decision makers or the entire firm aim to accomplish; delivery in a form of a clear, well-told story;

  - Data volume—specialists are needed to distill this data into manageable volumes of meaningful information
  - Delivery by convenient channels

- **Action**:

- – According to Forrester principal analyst Brian Hopkins (2016), firms are turning data into insight and action by building systems of insight—the business technology and discipline to harness insights and consistently turn data into action. ... Connecting data to action through insight development in a closed loop is critical for business. (We may note that this leads to cyclical nature of BI, discussed in the next chapter.) To discover insights, you don't need more data; you need systems of insight driven from the top by a culture that values insights-to-execution.
- – Insights are the missing link between data and action. According to (Dykes, 2016), key attributes of an actionable insight are: alignment—insight metrics are key to your business; context—additional data to explain or benchmark; relevance—delivered to the right person at the right time in the right setting; specificity—an insight is complete and well-focused; novelty—things that have been not known before; clarity—insight is clearly communicated and not messed-up.

- **Storage and reuse**: current insights might be as well useful in the future, but as their record creates additional work, its value should be promoted.

  - – Insight reuse is an important issue that closely relates to experience management and lessons learned.
  - – Personnel turnover is likely to have a negative effect that diminishes benefits brought by longevity and buildup of significant understanding of business. There should be a tradeoff between this significant understanding based on experience and stereotypes that impair change; this statement is also relevant for experience and lessons.

This sequence is rather reminiscent of decision support process, and obviously an important decision should be preceded by an insight.

While an insight is probably an ideal product of BI, not every BI product can be considered an insight. It might be a noticed important discrepancy leading to serious discoveries and earning or saving significant money. This triggering information, if proven right, will serve as a foundation for important insights in the area, not mentioning the possible side products. If we define an insight as a unit of integrated sense that required information integration in its production, then in the same way more complex, higher-power insights can be produced from simpler insights. Quoting Marchand and Peppard (2013), "While there may be some "aha" moments, when ideas and insights emerge quickly, there will be many more occasions when managers—and not just data specialists and analysts—must rethink the problem, challenge the data, and put aside their expectations."

References

Arnott, D., & Pervan, G. (2005). A critical analysis of decision support systems research. *Journal of Information Technology, 20*(2), 67–87.

Arnott, D., & Pervan, G. (2014). A critical analysis of decision support systems research revisited: The rise of design science. *Journal of Information Technology, 29,* 269–293.

Atre, S. (2004). *What is business intelligence? Interview with Tony Shaw.* Retrieved December 1, 2017, from http://dssresources.com/interviews/atre/atre07092004.html.

Boyer, J., Frank, B., Green, B., Harris, T., & Van De Vanter, K. (2010). *Business intelligence strategy: A practical guide for achieving BI excellence.* Ketchum, ID: MC Press Online.

Bucher, T., Gericke, A., & Sigg, S. (2009). Process-centric business intelligence. *Business Process Management Journal, 15*(3), 408–429.

Calof, J., & Wright, S. (2008). Competitive intelligence. *European Journal of Marketing, 42*(7/8), 717–725.

Chou, D. C., Tripuramallu, H. B., & Chou, A. Y. (2005). BI and ERP integration. *Information Management and Computer Security, 13*(5), 340–349.

Clark, T. D., Jones, M. C., & Armstrong, C. P. (2007). The dynamic structure of management support systems. *MIS Quarterly, 31*(3), 579–615.

Dalton, J. (2016). *What is insight? The 5 principles of insight definition.* Retrieved December 1, 2017, from https://thrivethinking.com/2016/03/28/what-is-insight-definition/.

Daly, M. (2016). Decision support: A matter of information supply and demand. *Journal of Decision Systems, 25*(S1), 216–227.

Davenport, T. (2005). *The right way to use business analytics.* Retrieved December 1, 2017, from http://www.cio.com/article/2448414/metrics/the-right-way-to-use-business-analytics.html.

Davenport, T. (2014). 10 Kinds of stories to tell with data. *Harvard Business Review*, May 5, 2014. Retrieved December 1, 2017, from https://hbr.org/2014/05/10-kinds-of-stories-to-tell-with-data.

Davenport, T. (2015). The insight-driven organization: Management of insights is the key. *Deloitte Insights,* August 26, 2015. Retrieved August 23, 2018, from https://www2.deloitte.com/insights/us/en/topics/analytics/insight-driven-organization-insight-management.html.

Davenport, T., & Harris, J. (2007). Competing on analytics. *Harvard Business Review, 84*(1), 99–107.

Dawson, L., & Van Belle, J. P. W. (2013). Critical success factors for business intelligence in the South African financial services sector. *SA Journal of Information Management, 15*(1), 1–12. https://doi.org/10.4102/sajim.v15i1.545.

Dykes, B. (2016). *Actionable insights: The missing link between data and business value.* Retrieved December 1, 2017, from https://www.forbes.com/sites/brentdykes/2016/04/26/actionable-insights-the-missing-link-between-data-and-business-value/#48e0ef2251e5.

Elbashir, M. Z., Collier, P. A., & Davern, M. J. (2008). Measuring the effects of business intelligence systems: The relationship between business process and organizational performance. *International Journal of Accounting Systems, 9,* 135–153.

Evelson, B. (2010). *Want to know what Forrester's lead data analysts are thinking about BI and the data domain?* Retrieved December 5, 2017, from https://go.forrester.com/blogs/want-to-know-what-forresters-lead-data-analysts-are-thinking-about-bi-and-the-data-domain/.

Feldman, M. S., & March, J. G. (1981). Information in organizations as signal and symbol. *Administrative Science Quarterly, 26*(2), 171–186.

FitzGerald, J., FitzGerald, A., & Stallings, W. (1981). *Fundamentals of systems analysis.* New York, NY: John Wiley & Sons.

Gartner Business Intelligence Glossary. (2017). Retrieved June 12, 2018, from http://www.gartner.com/it-glossary/business-intelligence-bi/.

Gilad, B. (2004). *Early warning: using competitive intelligence to anticipate market shifts, control risk, and create powerful strategies.* New York, NY: Amacom Books.

Gremillion, L., & Pyburn, P. (1988). *Computers and information systems in business*. New York, NY: McGraw-Hill.

Holsapple, C. W., & Sena, M. P. (2005). ERP plans and decision-support benefits. *Decision Support Systems, 38*, 575–590.

Hopkins, B. (2016). *Think you want to be "data-driven"? Insight is the new data*. Retrieved December 1, 2017, from https://go.forrester.com/blogs/16-03-09-think_you_want_to_be_data_driven_insight_is_the_new_data/.

Imhoff, C., & White, C. (2008) *Full circle: Decision intelligence (DSS 2.0)*. Retrieved December 5, 2017, from http://www.b-eye-network.com/view/8385.

Kandogan, E., Balakrishnan, A., Haber, E. M., & Pierce, J. S. (2014). From data to insight: Work practices of analysts in the enterprise. *IEEE Computer Graphics and Applications, 34*(5), 42–50. https://doi.org/10.1109/MCG.2014.62.

Kobielus, J. (2010). *What's not BI? Oh, don't get me started … Oops too late … Here goes….* Retrieved December 1, 2017, from https://go.forrester.com/blogs/10-04-30-whats_not_bi_oh_dont_get_me_startedoops_too_latehere_goes/.

Kroenke, D., McElroy, M., Shuman, J., & Williams, M. (1986). *Business computer systems*. Santa Cruz, CA: Mitchell Publishing.

Laudon, K. C., & Laudon, J. P. (2006). *Management information systems: Managing the digital firm*. Harlow, UK: Pearson Education.

Lönnqvist, A., & Pirttimäki, V. (2006). The measurement of business intelligence. *Information Systems Management, 23*(1), 32–40.

Marchand D. A., & Peppard J. (2013). Why IT fumbles analytics. *Harvard Business Review*, January–February 2013.

McCrary, T. (2011). *3 Top sources of strategic marketing insights*. Retrieved August 20, 2018, from https://www.chiefoutsiders.com/blog/bid/68693/3-top-sources-of-strategic-marketing-insights.

Meredith, R., Remington, S., O'Donnell, P., & Sharma, N. (2012). Organisational transformation through Business Intelligence: Theory, the vendor perspective and a research agenda. *Journal of Decision Systems, 21*(3), 187–201.

Negash, S., & Gray, P. (2003). Business intelligence. *Proceedings of AMCIS 2003 (Americas Conference on Information Systems)*. pp. 3190–3199.

Pick, J. (1986). *Computer systems in business*. Boston, MA: PWS Publishers.

Popovič, A., Hackney, R., Coelho, P. S., & Jaklič, J. (2012). Towards business intelligence systems success: Effects of maturity and culture on analytical decision making. *Decision Support Systems, 54*, 729–739.

Popovic, A., Turk, T., & Jaklic, J. (2010). Conceptual model of business value of business intelligence systems. *Management, 15*, 5–30.

Porter, M. (1980). *Competitive strategy*. New York, NY: Free Press.

Power, D. (2013). *Decision support, analytics, and business intelligence* (2nd ed.). New York, NY: Business Expert Press, LLC..

Ramakrishnan, T., Jones, M. C., & Sidorova, A. (2012). Factors influencing business intelligence (BI) data collection strategies: An empirical investigation. *Decision Support Systems, 52*, 486–496.

Riabacke, A., Larsson, A., & Danielson, M. (2011). Business intelligence as decision support in business processes. *Proceedings of 2nd International Conference on Information Management and Evaluation (ISIME)*. pp. 384–392.

Rouibah, K., & Ould-ali, S. (2002). PUZZLE: A concept and prototype for linking business intelligence to business strategy. *Journal of Strategic Information Systems, 11*(2), 133–152.

Sabherwal, R., & Fernandez, I. (2011). *Business intelligence: Practices, technologies, and management*. Hoboken, NJ: Wiley.

Sharda, R., Delen, D., & Turban, E. (2014). *Business intelligence. A managerial perspective on analytics*. Upper Saddle River, NJ: Pearson Education.

Skyrius, R. (2004). Matching information technology services and needs. *The Global Business and Finance Research Conference.* London, July 2004.

Skyrius, R. (2008). The current state of decision support in Lithuanian business. *Information Research, 13*(2), 345. Retrieved December 10, 2017, from http://InformationR.net/ir/31-2/paper345.html.

Skyrius, R., & Bujauskas, V. (2011). Business intelligence and competitive intelligence: Separate activities or parts of integrated process? *Proceedings of the Global Business, Economics and Finance Research Conference.* London, July 2011.

Top Business Intelligence Trends: What 2800 BI Professionals Really Think. (2017). *Interactive.* Retrieved December 1, 2017, from https://bi-survey.com/top-business-intelligence-trends-2017.

Tyson, K. (2006). *The complete guide to competitive intelligence.* Chicago, IL: Leading Edge Publications.

Vriens, M., & Verhulst, R. (2008). Unleashing hidden insights—Business insights need to be extracted and prepared for impact. *Marketing Research, 20*(4), 12–17.

Vuori, V. (2006). *Methods of Defining Business Information Needs. Frontiers of e-Business research conference.* Tampere, Finland: Tampere University of Technology.

Watson, H., & Wixom, B. (2007). The current state of business intelligence. *Computer, 40*(9), 96–99.

Wills, S. (2013). Insight management. A new profession. Insight management academy. Retrieved December 1, 2017, from https://www.insight-management.org/sites/insightmanagement.org/files/2013-09-the-insight-management-profession.pdf.

Wixom, B., & Watson, H. (2010). The BI-based organization. *International Journal of Business Intelligence Research, 1*(1), 13–28.

Ziama, A., & Kasher, J. (2004). *A data mining primer for the data warehousing professional.* Retrieved December 1, 2017, from https://academics.teradata.com/Library/Samples/A-Data-Mining-Primer-for-the-Data-Warehouse-P.

Chapter 3
Business Intelligence Information Needs: Related Systems and Activities

3.1 Information Needs

In the previous chapter it has been stated that BI aims to produce insights to develop and maintain an advanced level of informing. However, the information needs that BI is intended to serve are not uniform and deserve a closer look at how these needs are met. Previous research on business information needs by the author has rounded up several prominent dimensions of business information needs:

- Simple to complex;
- Common to special;
- Repetitive to non-repetitive, or random.

We can define the principal differences between simple and complex information needs as follows (Skyrius & Bujauskas, 2010):
Simple Needs:

- Are of routine and, more often than not, repeating nature, based on simple questions leading to routine actions or simple decisions;
- They employ data from a single source or a small number of sources that are easily accessible;
- Few procedures to produce results that are mostly controlled by single own information system;
- Results are a direct pre-planned product or simple by-product of existing information system.

Complex Needs:

- Are much less of a routine nature;
- Are initiated by important multi-faceted issues;
- Often based on vaguely structured questions;
- Questions may change over time;

© Springer Nature Switzerland AG 2021
R. Skyrius, *Business Intelligence*, Progress in IS,
https://doi.org/10.1007/978-3-030-67032-0_3

- Require composite results drawn from data sources that are numerous, not exactly compatible, eclectic and often external;
- Use heterogeneous and recursive procedures;
- There is an increased role of soft information and judgment;
- The information sources and conditions of their use are not always controlled by own information system;
- Such needs cannot be exactly estimated beforehand and are hard to plan.

Complexity of information needs, as it is understood here, can be approximately estimated by evaluating features like:

- Number of information sources to be used, together with their quality and reliability;
- Number of procedures and stages required to produce the result;
- Number of participants involved;
- Number of dimensions to be considered—business, technical, social, political, environmental etc.;
- Rate of change in task goals that may introduce changes in goals, procedures and dimensions.

Of course, complexity because of its vagueness is not easily measurable by approaches like the one above. On the other hand, situations might be encountered that force to evaluate complexity—e.g., a need to produce a documented record of decision making to solve a complicated problem. In such case, either standard recording procedures that have been developed over time will be in place, or the information managers will have to decide upon what are the important features of the problem and decision to be preserved in a record. We may assume that the number of such features would roughly reflect the complexity of the problem, decision and related information needs.

In a vast variety of business information needs, the needs served by BI mostly belong to the complex part of the "simple-complex" dimension. This is because of at least higher-than-average sophistication that differentiates BI as a function which cannot be provided just by any information system. This sophistication can be reflected in many ways, and some of them have been discussed in the authors' earlier work related to information needs (Skyrius, 2008; Skyrius & Bujauskas, 2010) with dimensions like simple/complex and common/special needs.

Complex information needs reflect ongoing or emerging concerns, often in some kind of problem situations requiring important decisions and experiencing significant potential risks (e.g., deteriorating market potential, market crash, aggressive move by competitors, natural calamity) or significant rewards (e.g., market niche, significant innovation, new business model). Several typical examples of situations with complex information needs would be:

- If a worsening situation is recognized—e.g., increasing flow of customer complaints, what forces have caused this worsening, and what measures would lead to what outcomes?

- What is the estimation of possible risk associated with new competitors or substitute products?
- In a public procurement deal, what is the most objective set of proposal evaluation criteria?
- In an acquisition deal, what does the buying side need to know before the decision of go/no-go (data rooms, public information, due diligence, own sources, consultant advice etc.)?
- What is the nature and severity of the emergency situation?

The multifaceted nature of complex needs suggests some dimensions for their possible classification, with subsequent differences in satisfaction approaches (Skyrius & Bujauskas, 2010):

- **Urgency.** One extremity is emergency situations that need quick yet well-grounded decisions in a fast-changing environment; the other pole is non-emergency situations—e.g., acquisition, privatization, strategic foresights. Support-wise, urgent needs should require the presence of emergency informing systems.
- **Coverage.** On one hand, there are situations with wide scope, affecting the whole organization; on the other hand—a narrow yet complicated problem area; in the latter case we deal with reduced set of dimensions, and at the same time the need to go down into the "information silo" of the narrow area.
- **Required precision.** In some cases, a rough estimate of a situation is sufficient; in other cases, exact or near-exact results are required. For rough estimates there are "quick fix" models; for accurate calculations and estimates issues of source information quality and reliability come up. This dimension is related to urgency in a sense that urgent situations usually do not allow for time-consuming thorough estimates and have to deal with quick and rough assessments.
- **Heterogeneity,** defined by many or few information sources, procedures or participants. Highly heterogeneous needs would require information integration mechanisms for source heterogeneity; unified or transparent environments for procedure integration; convenient communication channels for participant input integration. This dimension is related to coverage in a way that problems of a wide coverage invoke use of a number of information sources, procedures and participants.
- **Structuredness.** Although complex information needs by their nature are on the unstructured side, variations are possible in a sense that some problems possess more structure than others—e.g., in the area of company mergers and acquisitions the principal set of procedures, although rather complex, is known beforehand; the launching of an innovative business model which had not existed before is considerably less structured.
- **Associated risks,** defined by size and probability of possible loss if incorrectly assessed. The high-risk situations require the use of risk-estimating procedures and evaluation of different scenarios; information triangulation and cross-checking for reliability of multiple sources might be used for increased reliability of results.

Table 3.1 Relation of simple-complex and common-special needs

	Simple needs	Complex needs
Special needs (problem-specific)	Importers of article X Legal conditions for business in a new market	Our potential in a new market Possible actions by competition
Common needs (available permanently)	Cash-at-hand Today's completed orders Fully serviced customer complaints Levels of inventory	Financial ratios Market share Benchmarking against competition

Source: author

In several well-known models of information activities, namely, Kuhlthau's ISP (information search process) model (Kuhlthau, Heinström, & Todd, 2008), Wilson's model of information behavior (Wilson, 2000), Cheuk's information seeking and use process model (Cheuk, 1999), the shift from initial detection of a problem situation to a problem-specific focused search and analysis reflects a transition from one type of information needs to another. If a common overview of business environment detects an issue worth attention, intelligence resources are centered on this specific issue. With respect to this transition, one more dimension of information needs can be introduced regarding specific attention for a certain situation. The routine, repeating needs can be named common needs, and specific, non-routine needs that zoom in on a certain problem are special needs. A similar dimension has been introduced by Gill (2015), containing both routine needs like monitoring of the environment, and non-routine needs like decision making and problem solving.

Common needs are of repeating nature and their satisfaction can be planned beforehand; the satisfaction procedures are clear, tested and reusable; all important areas of activity are given roughly the same attention; they closely relate to Choo's term "environmental scanning" (Choo, 2002). Common needs mostly relate to monitoring, signal detection and interpretation using repeatable, mostly tried-and-tested procedures; a portfolio approach seems suitable for distribution of attention and required resources.

Special needs emerge for a specific situation; they require extra attention and analysis; are of semi-structured and unstructured nature, random and hard to plan; the reuse of their procedures because of their random or unique nature is limited. Special needs, as opposed to a wide scope of common needs, center on a single (usually) issue of interest, may employ a wide palette of analytical tools and have a narrow or deep scope.

Several examples of combinations of above needs are presented in Table 3.1.

Simple needs do not satisfy the need for specific insights, and are used more for monitoring of environment and detection of issues. It is the complex and special part of the above-named dimensions that requires moving towards clarity and sense-making. The efforts to achieve required clarity are expected to fill gaps in current understanding of a problem. It is not just the collection of additional information, but

Table 3.2 Relation between common-special and repetitive—non-repetitive needs

	Repetitive	Non-repetitive
Special	**Special repetitive**: non-routine, but come up from time to time; narrow focus on an object or a group of objects, but the problems are of recurring nature and a known framework is used; an example situation can be mergers and acquisitions. The potential of IT support is medium to high.	**Special non-repetitive**: focus on a singular and little-researched object or problem; information needs are not fully defined and keep changing; an example can be a development of a new business model. IT support is limited and provided in an interactive way using standard tools, possibly in rather innovative ways.
Common	**Common repetitive**: routine repeating needs are fulfilled using known procedures, tools and application framework; information needs are known in advance; an example can be today's completed orders. The potential of IT support is high.	**Common non-repetitive**: of unique or random nature; encompass a class or group of objects under some non-routine circumstances; it can be a new market, new field of activity or market innovation; application of traditional tools and techniques may prove inadequate—e.g., the attention has been focused in wrong direction; an example can be monitoring of emerging innovations. The potential of IT use is low to medium.

Source: author

directed set of actions (Choo, 2002) or separation of signal from noise to solve contradictions and achieve sufficient understanding of a problem at hand.

One more possible dimension to be considered here is **repetitive** and **non-repetitive** information needs. Though seeming to have much in common to the dimension of **common-special** needs, it can be treated as a separate dimension: if both dimensions are combined in the above fashion, there are possible examples in all four quadrants, as shown in Table 3.2:

This comparison of two dimensions provides insight into the potential of IT support for different needs. Obviously, the segment of **common repetitive** needs has the highest potential for "softwarization". Straub and DelGiudice have named the information activities of this segment "plebian use of the system" (Straub & DelGiudice, 2012). Same can be said about the segment of **special repetitive** needs: although the situations here are not exactly routine, the same set of procedures can be applied each time with little or no customization. As an example, the field of corporate mergers and acquisitions falls into this category, and standard due diligence procedures, though not an everyday issue, are performed along the same requirements. The segment of **common non-repetitive** needs is somewhat controversial; it might exist as, for example, one-time inquiry into a field or a wide set of objects, but it has to be agreed whether it can be considered a common need if the inquiry is centered around a specific feature or group of features. Because of non-repetitive nature reusable IT solutions are hard to develop, so not much more than a loose and flexible set of IT tools and techniques can be expected to provide support, and their use most likely will be iterative. Same approach applies to the

Table 3.3 The number of decision cases with known information needs (Skyrius, 2008)

Market information—customers, sales, needs, opportunities	49	31%
Competition information—competitors' status, strength, intentions, actions	29	18%
Internal information—financials, capacity, inventory	27	17%
Legal information—laws, regulations, standards	26	16%
No such cases	26	16%
Technical information	2	1%
Total:	**159**	

segment of **special non-repetitive** needs, characterized by problem-specific use of standard tools and techniques like simulation, statistical analysis or pattern recognition. This segment addresses non-routine situations and decisions, so we can note that the definition "loose and flexible set of IT tools and techniques" sounds very much like the early definitions of decision support systems (DSS) technology.

The handling of complex information needs invokes a staged process where some information needs are known beforehand, and others are not—they emerge in the process of problem analysis or decision development. The existing research on such process recognizes its complicated nature. As stated by Choo (2016: 147), "The lack of information may be met by gathering more facts that are relevant to an issue, but the lack of clarity has to be met differently, by finding a frame of reference or testing hypotheses so that the organization can make sense of what is going on in the problematic situation." Information from various internal and external sources is needed to provide an assessment of how the problem features match past experience. In the case of contradiction, explanations are required that incur additional information needs. The contradictions or gaps in competence force to seek expertise or additional competence within and outside organization. In other words, clarity and sense-making require not just more information on the issue at hand, but a directed set of actions aiming at satisfactory sense-making, or "connecting the dots" to enable action.

In an earlier research on decision support information needs, performed by the author in the years 2004–2007 (Skyrius, 2008), part of the survey addressed issues on known and unknown information needs. The data from responses is presented below.

Have there been cases when information needs have been known before-hand? If so, what kind of information? The responses to this question have been grouped into the following groups (Table 3.3):

This group mostly includes information whose content and location are well known and generally accessible because of earlier experience in related cases. It covers a part of decision making information needs in the initial phase of decision making process, is easy to plan and prepare, and this information or its access points can be contained in close proximity to the decision makers.

Have there been cases when information needs have not been known before-hand, having emerged only while making a decision or too late altogether? If so,

Table 3.4 The number of decision cases with unexpected information needs (Skyrius, 2008)

No such cases	86	38%
Yes, there have (without specifying the information)	46	20%
Market information	23	10%
Internal information	15	7%
Competition information	14	6%
Legal information	14	6%
Technical information	14	6%
Informal, "soft" information—opinions, foresights	12	5%
Confidential information—e.g., reliability checks	5	2%
Total:	**229**	

what kind of information? The responses to this question have been grouped into the following groups (Table 3.4):

This group of responses can be explained by complexity of the problem, as well as changing conditions around the problem. This information is hard to plan and cannot be kept prepared and handy; instead, some generic information sources that are most often used in such situations can be made ready to use whenever required, including electronic public archives, search engines, directories etc. Same types of information are present in both groups, indicating that for a certain type of information, part of it is already available, while the development of an insight or decision might reveal additional needs for this information type. Several possible examples might be:

- Apart from our direct competitors and competing products, what are the possible sources of unconventional competing products and substitutes?
- What other sources of financing might be available for a project?
- Are there any legal acts that impose additional regulations on certain activities?

This paragraph attempts to reflect, at least in part, the existing variety of business information needs by their coverage and nature of emergence. The understanding of this variety should facilitate better understanding of business intelligence processes and technologies required to cover this variety, so the next paragraph addresses the process aspect of BI.

3.2   Business Intelligence Cycle

BI activities, like all informing activities, are of permanent and evolving nature. Research sources that discuss BI phases describe them with certain variations, but most if not all of them agree that BI process is of cyclical nature. Table 3.5 below presents several examples of the structure of BI process, based on several sources (Meyer, 1991, Herring, 1999, Choo, 2002, and Vuori, 2006, partially drawing from Choo).

Table 3.5 Examples of the structure of BI process

Herring (1999)	Meyer (1991)	Choo (2002)	Vuori (2006)
1. Identify key intelligence topics	1. Study raw data and information	1. Collection of data and information	1. Defining information needs
2. Create the knowledge base	2. Argue and debate what it means	2. Evaluation/filtering	2. Information gathering
3. Intelligence collection and reporting	3. Check and recheck facts	3. Storage	3. Information processing
4. Making the intelligence actionable and understandable	4. Resolve the inevitable inconsistencies in data	4. Analysis	4. Analysis
5. Dissemination to other users	5. Question original assumptions	5. Dissemination	5. Dissemination
	6. Interview experts		6. Utilization and feedback
	7. Develop theses		
	8. Test and retest		

Source: author

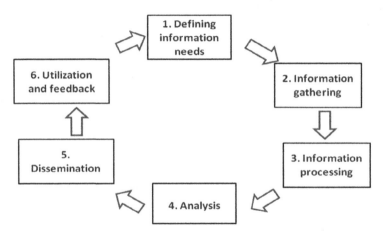

Fig. 3.1 A generic business intelligence process model (Vuori, 2006)

The cycle structure, as suggested by Vuori (Fig. 3.1, from (Vuori, 2006)), ensures that the received feedback helps to reevaluate or redefine information needs, if inconsistencies are discovered, and to create a body of experience as a side-product. The model depicts a generalized intelligence cycle whose elements roughly correspond to other suggested models, but the significant addition is the explicit feedback phase.

While all stages of this loop are important and every one of them may become the weakest link, the initial stage 1—"Defining information needs" serves as a foundation for subsequent activities. The first step of the intelligence cycle is supposed to start with the user expression of their information needs to information managers. According to Hulnick (2006), this does not exactly work for several reasons. Firstly, the communication to deliver guidance does not always work as expected—users often assume that the information (intelligence) system will alert them to problems, or provide judgments about the future. Often information managers have to take the initiative to obtain the information on intelligence needs. Even then, their requests are sometimes ignored. Secondly, information needs are often driven by current environment. This is supposed to work well with known knowns and known unknowns, but risks losing relevance in a changing environment. Information managers often know the gaps in intelligence data, but this not always the case.

As well, the overall structure of the cycle model has received criticism from intelligence practitioners and researchers. Treverton, Thomson, & Leone, 2001 (in (Tropotei, 2018)) have suggested that a cyclical model should be enhanced with interrelations between stages, reflecting the often-iterative nature of intelligence process.

Stage 2 covers information collection procedures. A portfolio principle is often applied to spread information collection resources over available information sources to achieve the best mix. Meige and Schmitt (2015), discussing intelligence cycle, have noted that in data collection, the narrower the scope, the less likely that environment monitoring will detect important issues.

Both stages 3 and 4 produce intelligence product from collected information: stage 3 performs the technical procedures on collected information—organizing, grouping, quality control, while stage 4 performs analysis and sense making with joint use of human and IT intelligent resources.

Stage 5 disseminates the analysis results among decision makers and other users for whom the intelligence product is intended. At this stage, culture—organization culture, information culture and intelligence culture largely defines the dissemination map. More issues regarding information culture and intelligence culture are discusses in Chap. 9 of this book.

Stage 6 is the point at which the information needs of the intelligence end users are to be satisfied. It also evaluates how effective the intelligence process is, and provides feedback on possible improvements, if necessary.

As in many cyclical process structures for some kind of informing activities, the importance of feedback step is evident—it closes the learning loop, creating experience and lessons that can be reused directly for similar cases or indirectly through enriched problem-solving skills and heuristics.

3.3 Business Intelligence and Related Informing Activities

As BI interacts with multiple other information systems and environments, this paragraph aims at determining/resolving the position of BI function regarding other informing activities within an organization. The separation of BI as a distinct type of information systems implies BI positioning in the set of information systems that are related to BI:

- **LOCATION aspect**—location of BI regarding other systems in the information value chain: management information systems (MIS), enterprise resource planning (ERP), business process management (BPM), e-business. These systems are for large part operations-level systems that provide a significant part of raw input to BI systems.
- **USER aspect**—differences of BI users differ from users of other systems. BI users are decision makers, strategy developers, process and function managers that use BI to support their actions and decision with consequences from short term to long term. Although informed actions are required at all management levels, BI concentrates on analysts and decision makers of middle and higher levels, thus raising the stakes—both the expected payoffs and the price of error are high.
- **FUNCTION aspect**—how BI informing activities differ from other systems. Other intelligent systems—decision support systems (DSS), artificial intelligence (AI), knowledge management (KM) relate to BI by research aspect, and by serving complex information needs for advanced informing. Entrepreneurship and innovation activities use BI for directions (e.g., to understand the behavior of competition and evaluate the need to react) and idea generation when innovative ideas can be generated out of available cues.
- **SOURCES aspect**—BI handles a much larger number of sources for data and information from the variety of formal and informal systems, in a variety of types and standards, and sometimes in huge volumes.

The remainder of this chapter covers all of the above aspects, starting from positioning BI in the information value chain, and continuing with other aspects that accentuate specific features of BI activities.

3.3.1 Business Intelligence and Management Information Systems

Figure 3.2 depicts how the various types of management information systems are positioned regarding business intelligence systems and activities in an organization.

Fig. 3.2 Positioning of BI against other information systems

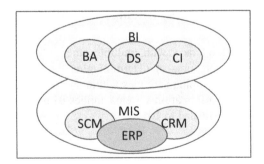

Here the lower group includes the common types of MIS—ERP (Enterprise Resource Planning), SCM (Supply Chain Management) and CRM (Customer Relationship Management) systems that serve the principal value chain and collect the basic body of facts taking place in organization's activities. The upper group joins the information systems and activities operating in the interests of business intelligence: BA (Business Analytics), DS (Decision Support) and CI (Competitive Intelligence).

Since the introduction of first information systems in business and management, the user expectations have been driven quite high—the systems were meant to provide reliable up-to-date information about all aspects of activities, point to opportunities and threats, support decision making and many other things by properly informing users. The timeline of developments in information systems shows emergence of new kinds of systems as information technology and user skills gain power. Some systems, like transaction processing systems and business process management, stay until today, and some, like executive support systems, do not.

The regular management information systems, intended to serve current information needs, proved their usefulness in day-to-day management tasks like accounting or management of sales and inventory. However, their inertia in handling non-routine informing tasks in semi-structured or unstructured situations led to development of decision support systems that were intended to handle the complex part of management information needs. Eventually, decision support function had merged with business intelligence activities covering both permanent monitoring of internal and external environment, and zooming in on specific problems requiring specific decisions.

Both management information systems and business intelligence systems maintain the principal function of providing the users with required meaningful information at required time and in required place. Same can be said about the role of any information system, but in the case of business intelligence we usually consider information needs that range in their complexity levels from medium to high, with correspondingly matching levels of sense making. Yet another important feature of business intelligence is multidimensionality, which extends from basic figures and facts to meaning-rich contexts, from monitoring of everyday operations to the analysis of large one-of-a-kind problems, from internal efficiency to wide-ranging insights at macro level. This provision of multiple dimensions and heterogeneous views, as opposed to MIS/ERP systems, leads to a more complete understanding about activities, opportunities and risks, and thus brings at least a part of the spectrum of information demands served by business intelligence into the area regarded in the author's earlier work as complex information needs. Advanced informing and complex information needs have many things in common, e.g. a required result might be a simple piece of information or a yes/no case, but the preceding distillation of multitude of aspects related to a problem clearly places it in the complex part of the spectrum of information needs.

According to Herring (1999), people engaged in intelligence functions focus on the task of identifying and defining management needs that actually require intelligence and not just information that could be acquired from organization's own resources or public sources. Intelligence here is understood as insight development based on collection of key information and performed in a systematic way. B. Liautaud, the founder of Business Objects, states that "intelligence elevates information to a higher level within an organization", and that it "contributes to an organizational state that may be characterized as collective intelligence" (Liautaud & Hammond, 2001). Therefore, simple informing alone cannot be considered business intelligence, although it can serve an important part in the intelligence information value chain. Negash and Gray (2003) support this view by pointing out that ordinary reporting is not adequate to satisfy comprehensive management information needs. According to Carton (Carton, 2007), several more important aspects of relations between BI and ERP systems are:

– Decision support required by managers is (mostly) aggregate operational data from ERP-type systems;
– BI tools are used to "bridge the gap" between operational systems and reality;
– Data handling by BI-related procedures like ETL (Extract, Transform, Load) re-introduces data integrity and timeliness issues.

The problems of MIS capability lie in their limited scope—they work only on data that has been recorded and uploaded to the system, predominantly in a structured format. Therefore, their potential in handling heterogeneous data, broad contexts, hidden interdependencies and cues is limited. However, the role of everyday "workhorse" systems like ERP in assuring effective BI is somewhat underrated; meanwhile, they have huge influence on data quality, information quality and subsequent intelligence quality. Besides, these systems feed a significant part of complex information needs by properly doing their part on internal data. Of course, BI is expected to perform complicated and intelligent tasks, but recent developments in the business of BI tools (e.g., Salesforce Wave), as well as former research by the author (Skyrius & Winer, 2000) indicate that often simple tasks and functions, complementing each other, create the initial basis for good intelligence.

3.3.2 Business Intelligence and Decision Support

Decision support is not a clearly expressed corporate function like finance or human resources; one can hardly find a VP for decision support. Yet decisions are made every day at all levels of management. This paragraph is more about BI relation to DS function than to DS systems, which after their prominence in the last decade of XX century faded or morphed into BI.

The timeline of developments in the IS field usually shows (Arnott & Pervan, 2014; Power, 2013) the coming of DS and BI, the former happening significantly earlier—in the 70s–80s. In most if not all timelines, decision support morphs into business intelligence. From the time perspective, both types of systems have been created with an intent to serve complex information needs of both common and special types. There is a considerable difference between BI and DS in that while business intelligence is expected to cover a wide need for awareness throughout the entire business horizon, decision support concentrates on a certain problem. In other words, BI aims mostly at common information needs, while DS covers special information regarding a certain problem. We can assume that higher quality of BI informing leads to less support required at the time of facing a decision.

In management decision making, the complexity of the problem to be solved directly translates to the complexity of the information tasks to produce a well-supported decision. Such complexity creates substantial demands for support environment in terms of variety of information sources and incompatibility of information obtained from these sources; use of sophisticated and problem-specific analysis and modeling tools; use of communication and group support environments to fine-tune the interests of stakeholders, to name a few (Power, 2013). Successful management decision support applications boost intelligent management activities and improve decision quality, where the expected payoff is quite high. So it is no surprise that for a long time developers of decision support technologies have designed support products ranging from a simple functional support to fully automated decision making functionality, with a huge spectrum of analytical tools in between. Microsoft Excel spreadsheet is the most prominent example of the former (Bhargava, Sridhar, & Herrick, 1999), while expert systems in the nineties of XX century and contemporary automated systems for securities trading are examples of the latter.

Decision activities, apart from their direct intent to develop a solution to a problem, create valuable experience and contribute to refinement of problem-solving skills. The presence of feedback defines a cyclical nature of decision making activities. A well-known cyclical model of decision making is the 'OODA' loop (Observe, Orient, Decide, Act) that originates from military activity and has been proposed by Colonel John Boyd of the US Army (Richards, 2004). The cycle combines the stages of observing, perceiving, decision making, and taking action. Apart from these natural stages of the decision-making process, the necessary condition is to conceal one's OODA logic from the opponent, thus seeking to create the opportunity to recognize the phase of the opponent's OODA logic and utilize this for one's own purposes.

Following the decision cycle approach, a decision making process, based on (Skyrius, 2016), can be presented as shown in Fig. 3.3: an user-centered cycle surrounded by support environment.

The advent of BI has revived interest in decision support and the relation between the domains of BI and DS. In published research, different and sometimes opposing opinions on this relation can be found. For example, one of the notable authorities in the field of decision support, Daniel Power (2013) states that decision support encompasses business intelligence; on the other hand, numerous sources indicate that decision support is a part of BI activities (Arnott & Pervan, 2014; also V. Sauter, 2010) The author of this book proposes an interlocking 2-cycle model (Fig. 3.4), where the cycle 1 covers the business intelligence part, and in the case of detection of important issues requiring decisions, cycle 2 is activated. It can be stated that the *constant* state of adequate informing (the function of BI) improves the decisions that

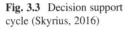**Fig. 3.3** Decision support
cycle (Skyrius, 2016)

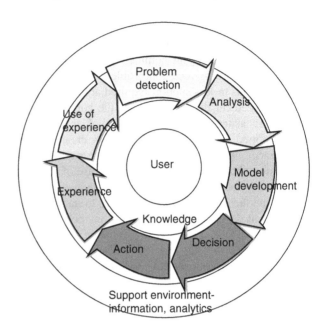

have to be made *occasionally*, raising chances to avoid surprises and stay proactive as possible.

This separation of BI and DS into two cycles does not necessarily mean a clear separation in informing activities, although they serve different purposes. According to Fleisher (2009), analysis tools should not be deemed equal to decision support tools. While the latter are used by individuals or groups to arrive at decisions, the former provide evolving conceptual frameworks on a growing body of knowledge. These tools are often combined with other factors—data, intuition, experience, judgment, and their results are recommendations and insights that might suggest action. In the double cycle model shown here, the adequately sophisticated environment scanning does produce insights of its own that leave less problems in decision making when a related problem requiring decision comes up. An appropriate BI tool should not produce more information, but instead add structure and clarity helping decision makers to achieve better decisions.

3.3.3 Business Intelligence and Competitive Intelligence

Competitive intelligence (CI), a vital part of business intelligence activities, has a task to cover a very important part of information needs spectrum—monitoring current and potential competitors. There are other types of external objects of interest—customers, suppliers, partners—that require monitoring. However, a business usually has direct contacts with such objects, often using integrated

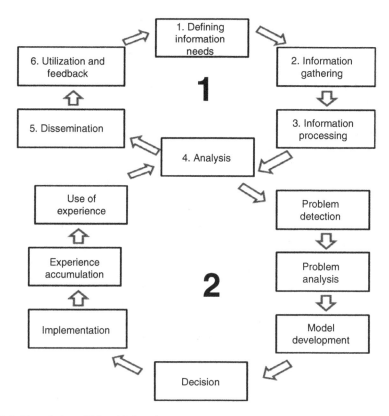

Fig. 3.4 The relation of BI and DS cycles

information activities and systems, which cannot be said about the competition, with rather rare exceptions.

One of the important dimensions for different approaches to relation between BI and CI that comes to attention is separation of business intelligence and competitive intelligence as being respectively directed inwards and outwards. Some sources (e.g., Nash, 2010) do state that business intelligence is a well-defined activity with tools and methods for estimating internal efficiency and possible subsequent automation of activity management—approaches like business process management (BPM) or key performance indicators (KPI) are frequently used in this context. Other sources (Combs, 1993) define competitive intelligence as being another well-defined standalone activity, the goals of which are to select, collect, interpret and distribute publicly-held information that has strategic importance. CI is an activity that mostly relies on external sources not covered by own information systems, and is much less structured than internal systems like CRM or SCM. Regarding the relation between internal and external BI coverage, Negash and Gray (2003) note that many firms still view BI as an inward-looking function and label the CI part of intelligence activities "outward-looking", accentuating the necessity for effective use of external information sources. According to Choo (1999), there is a hierarchy between intelligence functions that covers the relation between BI and CI: *competitor intelligence* is a part of *competitive intelligence*, which is a part of *business intelligence*, which is a part of *environmental scanning*.

In author's opinion, the division of business intelligence and competitive intelligence into their respective internal and external areas of interest is in contradiction with the very meaning of the term "intelligence". Intelligence presumes the level of understanding that is comprehensive and supported by possession of varied, multidimensional and mutually supportive information. In other words, BI has some specific features of at least higher-than-average sophistication that differentiate BI as a function that cannot be provided just by any information system. This sophistication can be reflected in many ways, and some of them have been discussed in the authors' earlier work related to information needs (Skyrius, 2008; Skyrius & Bujauskas, 2010) with dimensions like simple/complex and common/special needs. The boundaries of BI information needs include both routine needs like monitoring of the environment, and non-routine needs like decision making and problem solving (Gill, 2015). Simple and common needs can be mostly attributed to *routine* needs, while complex and special needs are attributed to *non-routine* needs, although this is not always the case—combinations like simple and special needs (e.g., contact information for a possible partner) or common and complex needs (e.g., market share or benchmarking against competition) do exist as well.

In the case of the emergence of information need that is recognized as complex, its complicated structure often encompasses both internal and external information needs. The problem issue can be quite internal, like rising production costs or impaired quality, but a solution to such problem shall usually need consideration of external conditions, benchmarks, resources and other factors. Similarly, the case of external issue requiring solution and creating special information needs most probably will depend on internal factors as well, such as policy issues or resources available.

CI methods, while entirely legal, borrow significantly from political or military intelligence methods. Many former intelligence employees have continued their career paths in competitive intelligence positions in business companies (Ben Gilad, Jan P. Herring), and there are influential professional societies uniting competitive intelligence professionals—e.g., Society of Competitive Intelligence Professionals, founded in United States in 1986, which later morphed into Strategic & Competitive Intelligence Professionals. The CI professional's activities should follow the ethical guidelines set by professional bodies and own organization.

Probably one of the most precise definitions of CI has been provided by Tyson (2006), who defined competitive intelligence as a systematic process that converts bits and pieces of competition information into strategic knowledge for decision making: knowledge about current competitors' position, historical performance, strengths and weaknesses as well as specific future intentions. According to the same source, intelligence objectives are:

- To avoid surprises;

- To identify threats and opportunities;
- To gain competitive advantage by decreasing reaction time;
- Improve long and short term planning.

As bits and pieces of valuable information on competition may come from a large variety of sources, the integration of this information becomes a key function in developing an adequate insight. Information integration is covered in more depth in Chap. 5 of this book. The need to integrate this scattered and sometimes controversial information requires not only advanced approaches in data engineering, but information sharing among people engaged in CI activities. According to Fleisher (2003), "CI works best in organizations that have information-sharing cultures. . . . every colleague must be involved as CI antennae". As the culture that supports information sharing is important for the entire set of BI activities, cultural issues are discussed in Chap. 9 of this book.

3.3.4 Business Intelligence and Business Analytics

The debate whether BI and BA are separate activities is apparently of academic nature. At the same time, a lot of opinions are driven by marketing trends and sales effort.

In a collection of opinions by business professionals (Heinze, 2016), Timo Elliott, a SAP Innovation evangelist, expresses a well-based opinion on differences between BI and BA that might be non-exist; "everybody has an opinion, but nobody knows, and you shouldn't care". In the same source, Rado Kotorov, a chief innovation officer at Information Builders, states that analytics is a subset of a broader platform of BI capabilities to access data, manage metadata, develop reports, publish and distribute packaged results. In difference from BA, BI includes selection of information sources, reliability evaluation, lessons learned, while BA is the production or transformation stage that turns data into information into insights.

There are other sources that propose separating BI from business analytics, and state that BI is looking backward, while business analytics look forward. Using this approach, intelligence is reduced to data collection, reporting and occasional development of insights, while analytics project the future, draw scenarios, are interactive and iterative. It is rather confusing to state that both approaches can be separated: according to this logic, BI does not check against our former projections and beliefs as reference points for evaluating current activities and events, and analytics can forecast without using past data. Yet other sources (e.g., Kandogan, Balakrishnan, Haber, & Pierce, 2014) use the term "Business intelligence analytics", and define BI as tools and processes to transform data into insights that support decision making. This definition implies that data is transformed into insights by analytical work that is performed mostly by business analysts. The same source brings up a term "analytical ecosystem", encompassing a variety of data sources, analytical tools, and participants for transforming data into insights. This definition of analytical work supports its role in satisfying complex information needs whose features, as presented earlier in this chapter, are very much in line with this definition.

On relation between BI and BA, in the same source (Heinze, 2016) Tim Biskuip of Progressive Business Publications, a business publisher, has presented a good rounding-up statement: "Analysis without intelligence can't be done … that's guessing and intuition (which still rely heavily on informal intelligence). Analysis that isn't intelligent is dumb. Capturing intelligence without doing analysis is a waste of time, and intelligence that isn't based on analysis of intelligence isn't intelligent."

3.4 Intelligence in Other Domains and Its Relation to BI

As mentioned earlier in this chapter, BI borrows much from other intelligence experiences, especially in the area of competitive intelligence. A number of other activities with needs for insights or discovery—state and political intelligence, areas with intensive analytics like scientific research or forensics—have been using methods for discovery or insight building that are now firmly adopted by BI. Because of the similarities, lessons learned in other domains are expected to be useful in business intelligence, especially regarding human factors. There are differences that have to be accounted for, but similarities in discovery techniques, information behavior and human factors are rather important. Just several arguments from published sources are presented below:

- Tropotei (2018), discussing intelligence cycle, has named BI a "secondary discipline", as compared to intelligence in general.
- Seth Grimes (2007), an industry analyst, noted when writing about the history of text mining: "The earliest text-mining users were investigators such as intelligence analysts and biomedical researchers looking for needles in haystacks: the terrorist, known or unknown, who might be detected through a pattern of actions and associations; a protein whose presence activates or inhibits a genetic pathway, leading to a cancer, that might be known by mining biomedical literature".
- Johnston (2005: xv): "I mention the surgical and astronautical studies for a number of reasons. Each serves as background for the study of intelligence analysts. Astronauts and surgeons have very high performance standards and low error rates. Both studies highlight other complex domains that are interested in improving their own professional performance. Both studies reveal the need to employ a variety of research methods to deal with complicated issues, and they suggest that *there are lessons to be learned from other domains*. Perhaps the most telling connection is that, because lives are at stake, surgeons and astronauts experience tremendous internal and external social pressure to avoid failure. The same often holds for intelligence analysts."
- An area long known for its intense work with discovery and sense making from large data volumes is astronomy and space research, where projects or clusters of projects like SETI (Search for Extra-terrestrial Intelligence) intake enormous

amounts of space monitoring data and comb it for anything that might resemble a signal from intelligent beings.

- Meige and Schmitt (2015), talking about innovation intelligence (monitoring of research and technology landscape for detection of important innovations), have noted the similarities between uses of intelligence in various activities—military, economic, competitive and so on. They note that "... regardless of the field of application, the essence of definition remains the same", i.e. support for insights and strategic decision making.

3.4.1 Similarities with Other Domains

The common features between business intelligence and other forms of intelligence are evident: there's an overarching need to have a balanced and complete view of the environment that prevents from surprises and supports proactive behavior. All of the above intelligence activities are expected to provide insights, delivering maximum sense and certainty. According to Johnston (2005), "analysis is a scientific process".

Issues regarding relation of BI to other kinds of intelligence:

- Similar goals—to reach for adequate level of informing and production of required insights as the principal intelligence product;
- Similar positioning of complexity of information needs;
- Multidimensionality, heterogeneity and dynamic nature of information sources, procedures, changing needs and requirements;
- Similar informing activities and analytical methods—collection of data and information from different sources, information integration, importance of analytical functions, etc.;
- Similar cultural issues—proper organizational culture, existence of intelligence community, information and insight sharing, repositories for experience and lessons learned;
- Frequent elements of novelty—dynamic environment and changing situations bring up issues that were not previously encountered;
- Because of dynamic environment, information behavior is subject to constant changes.

Such common traits mostly relate to the methods part of BI: the analysis should aim for insights and complete coverage; deal with assorted and chaotic information sources and types; support the avoidance of surprises. Bryce McCandless (2003; in Fleisher & Blenkhorn 2003) presented the seven principles for improving Marine Corps intelligence capabilities:

- Focusing on tactical intelligence—relevant, useful, and actionable for the organization;
- The emphasis must be downward—supported by senior management with recognition and resources;

- Intelligence drives operations—decisions should be made in wide encompassing context of internal and external issues;
- Intelligence must be directed and managed by multi-disciplined officers—in the case of business, these should be capable analysts with insight capabilities, not locked in data collection and number crunching;
- Products must be timely and individually tailored—relevance is still maintained, and usability is appropriate;
- Intelligence staffs use intelligence produced by intelligence organizations—intelligence developers should have a forum to present findings to key management people;
- Last step in the cycle must be utilization, not dissemination—directed delivery to actionable points/people.

Tim Biskuip (in (Heinze, 2016)) explicitly relates BI to the experience accumulated in political intelligence domain: "I have chosen to first look at Business Intelligence as it is the older term (least amount of controversy on that), it is also the only one that consistently shows up as being a real word and not needing to be auto-corrected. I am not referring to Intelligence as a definition of "smartness" . . . I am referring to Intelligence as a non-descriptive noun. . . . A tested definition of this exists in the CIA, the Central Intelligence Agency. The CIA is responsible for having the people and processes and infrastructure in place to capture data (contextual and numerical) from around the world, but also need to analyze, disseminate and strategize around how to intelligently apply the intelligence and output of the various analysis actions done to it. The intelligence teams exist to capture, analyze and strategize around the information (intel.)".

The processes of capturing and analyzing information are just one part of intelligence process, however; and there are prominent examples with bad political intelligence, preferably illustrating the limited impact of routine informing:

- The events of September 11, 2001 brought up what is now common knowledge about bad intelligence caused mostly by absence of sharing of important information between involved parties. The case goes down in history as an example of gross failure of major intelligence and security agencies to communicate and share intelligence information. A quote from the U.S. National Commission on Terrorism: "Earlier in this report we detailed various missed opportunities to thwart the 9/11 plot. Information was not shared, sometimes inadvertently or because of legal misunderstandings. Analysis was not pooled. Effective operations were not launched. Often the handoffs of information were lost across the divide separating the foreign and domestic agencies of the government. However the specific problems are labeled, we believe they are symptoms of the government's broader inability to adapt how it manages problems to the new challenges of the twenty-first century. The agencies are like a set of specialists in a hospital, each ordering tests, looking for symptoms, and prescribing medications. What is missing is the attending physician who makes sure they work as a team."
- Military campaign in Iraq in 2003 is another example of bad intelligence caused by political factors. (Peter Taylor, Daily Telegraph, 18 March 2013) "Much of the

key intelligence that was used to justify the war was based on fabrication, wishful thinking and lies—and as subsequent investigations showed, it was dramatically wrong. Saddam Hussein had no weapons of mass destruction (WMD)." ... "In the introduction to this dossier (known to some as the "dodgy" dossier) Tony Blair confidently declared, 'the assessed intelligence has established beyond doubt that Saddam has continued to produce chemical and biological weapons [and] that he continues in his efforts to develop nuclear weapons.'"

The above examples confirm the statement (Jervis, 2010) that political intelligence of all types of intelligence tends to be most influenced by policymakers, although BI and business decision support are not immune from political influence (Carlsson & Walden, 1995).

3.4.2   Differences from Other Domains

Although the similarities between BI and other types of intelligence are extensive, there are important differences that BI users have to be aware to prevent the possibly risky blind transfer of approaches from another domain. Below are several differences of intelligence in other domains from BI.

1. **State, military and political intelligence activities** are performed with substantial secrecy. According to Johnston (2005), there are "... huge internal and external pressures to avoid failure". Issues of misinforming and deception come up, and the users experience far greater risks dealing with possibly deceptive data than in business intelligence. In contrary to this, business intelligence presumes ethical informing activities.
2. **Law enforcement, or criminal intelligence**—as its activities are performed by state institutions, much required information on individuals and organizations is available by default on a state level. In BI case, such information would be deemed sensitive, of restricted use and therefore hard to access.
3. In **scientific research**, the issues under investigation are far less clear and difficult to measure. Compared to academia, BI features more accountability and evaluation of results. Muses (2012) stated that for business, so many iterations and summarizations in the process of sense making is a total luxury, as compared to academia. The demands of business environment are often rather harsh, especially regarding urgency and value of decisions involved. Sometimes one even knows the answer or the best alternative, and spends time finding the evidence.

Just like in any other case of dealing with experience from other domains, BI has to understand the potential value of applying this experience, as well as risks emerging from differences in core activities. Questions asked on data might be better defined in some domains like business or law protection (looking for known unknowns) and vaguer in others like scientific research (looking for unknown unknowns). Same can be said about time pressures that are essential in some domains but have little importance in other domains. There might be different priorities—e.g., according to Miller (2000), political intelligence pays prime attention to threats to national security, and less so on opportunities. In corporate environment risks are undoubtedly important, but opportunities take considerably more managers' time and effort.

The rounding-up statement of the chapter may be formulated as follows: BI, being understood as a set of methods, techniques and tools for advanced informing, is:

- Related to every other information activity and major information system in an organization; and
- Feeds off all other activities and systems that contain what may be considered elements of BI (components of intelligence process); also utilizes numerous extra-system activities and resources;
- Uses extensive intelligence experience gained in other intelligence fields with many similarities.

References

Arnott, D., & Pervan, G. (2014). A critical analysis of decision support systems research revisited: The rise of design science. *Journal of Information Technology, 29*, 269–293.

Bhargava, H., Sridhar, S., & Herrick, C. (1999). Beyond spreadsheets: Tools for building decision support systems. *IEEE Computer, 32*(3), 31–39.

Carlsson, C., & Walden, P. (1995). AHP in political group decisions: A study in the art of possibilities. *Interfaces, 25*(July–August), 14–29.

Carton, F. (2007). Studying the impact of ERP applications on managerial decision making. *IFIP 8.3 Task force meeting*, Cork, April 2007.

Cheuk, W. B. (1999, May 27). *The derivation of a "situational" information seeking and use process model in the workplace: Employing sense-making*. International Communication Association annual meeting, San Francisco, California. Retrieved February 27, 2006, from http://communication.sbs.ohio-state.edu/sense-making/meet/1999/meet99cheuk.html.

Choo, C. W. (1999, February/March). The art of scanning the environment. *Bulletin of the American Society for Information Science and Technology, 25*(3), 21–24.

Choo, C. W. (2002). *Information management for the intelligent organization: The art of scanning the environment*. Medford, NJ: Information Today.

Choo, C. W. (2016). *The inquiring organization*. New York, NY: Oxford University Press.

Combs, R. (1993). *The competitive intelligence handbook*. Chicago, IL: Richard Combs Associates Inc..

Fleisher, C. S. (2003). Should the field be called competitive intelligence? In C. S. Fleisher & D. L. Blenkhorn (Eds.), *Controversies in competitive intelligence: The enduring issues*. Westport, CT: Praeger Publishers.

Fleisher, C. S. (2009). The tools CI analysts use: An overview. In W. B. Ashton (Ed.), *Competitive technical intelligence*. Bonnie Hohhof: Competitive Intelligence Foundation.

Gill, T. G. (2015). *Informing science* (Vol. 1). Santa Rosa, CA: Informing Science Press.

Grimes, S. (2007, October 30). *A brief history of text analytics*. Business Intelligence Network.

Heinze, J. (2016). *Business intelligence vs business analytics: What's the difference?* Retrieved February 2, 2018, from https://www.betterbuys.com/bi/business-intelligence-vs-business-analytics/#comments.

Herring, J. (1999). Key intelligence topics: A process to identify and define intelligence needs. *Competitive Intelligence Review, 10*(2), 4–14.

Hulnick, A. S. (2006). What's wrong with the intelligence cycle? *Intelligence and National Security, 21*(6), 959–979.

Jervis, R. (2010). Why intelligence and policymakers clash. *Political Science Quarterly, 125*(2), 185–204.

Johnston, R. (2005). *Analytic culture in the U.S. intelligence community*. Washington, DC: Center for the Study of Intelligence.

Kandogan, E., Balakrishnan, A., Haber, E. M., & Pierce, J. S. (2014). From data to insight: Work practices of analysts in the enterprise. *IEEE Computer Graphics and Applications, 34*(5), 42–50. https://doi.org/10.1109/MCG.2014.62.

Kuhlthau, C. C., Heinström, J., & Todd, R. J. (2008). The 'information search process' revisited: Is the model still useful? *Information Research, 13*(4), paper 355. Retrieved December 1, 2017, from http://InformationR.net/ir/13-4/paper355.html.

Liautaud, B., & Hammond, M. (2001). *E-Business intelligence: Turning information into knowledge into profit*. New York, NY: McGraw-Hill.

McCandless, B. (2003). What key learning should corporate competitive intelligence specialists acquire from their military intelligence counterparts? In C. S. Fleisher & D. L. Blenkhorn (Eds.), *Controversies in competitive intelligence: The enduring issues*. Westport, CT: Praeger Publishers.

Meige, A., & Schmitt, J. (2015). *Innovation intelligence*. Philadelphia, PA: Absans Publishing.

Meyer, H. L. (1991). *Real-world intelligence*. Friday Harbour, WA: Storm King Press.

Miller, J. (2000). *Millenium intelligence. Understanding and conducting competitive intelligence in the digital age*. Medford, NJ: CyberAge Books.

Muses, R.S. (2012). *The sensemaking process and leverage points for analysts as identified through cognitive task analysis*. Retrieved June 15, 2019, from https://rachelshadoan.wordpress.com/2012/02/02/the-sensemaking-process-and-leverage-points-for-analyst-as-identified-through-cognitive-task-analysis/.

Nash, K. (2010). *Using BI, BPM data to change business processes fast. Computerworld, June 17, 2010*. Retrieved December 15, 2017, from http://www.computerworld.com/s/article/9178207/Using_BI_BPM_Data_to_Change_Business_Processes_Fast.

Negash, S, & Gray, P. (2003) Business Intelligence. *Proceedings of AMCIS 2003 (Americas Conference on Information Systems)*. pp. 3190–3199.

Power, D. (2013). *Decision support, analytics, and business intelligence* (2nd ed.). New York, NY: Business Expert Press, LLC.

Richards, C. (2004). *Certain to win*. Bloomington, IN: Xlibris, Inc.

Sauter, V. (2010). *Decision support systems for business intelligence*. Hoboken, NJ: John Wiley & Sons.

Skyrius, R. (2008). The current state of decision support in Lithuanian business. *Information Research, 13*(2), 345. Retrieved December 21, 2017, from http://InformationR.net/ir/31-2/paper345.html.

Skyrius, R. (2016). *Business information: Needs and satisfaction*. Santa Rosa, CA: Infoming Science Press.

Skyrius, R., & Bujauskas, V. (2010). A study on complex information needs in business activities. *Informing Science: The International Journal of an Emerging Transdiscipline, 13*, 1–13. Retrieved June 5, 2020, from http://inform.nu/Articles/Vol13/ISJv13p001-013Skyrius550.pdf.

Skyrius, R., & Winer, C.R. (2000). IT and management decision support in two different economies: A comparative study. *Challenges of Information Technology Management in the 21st Century: 2000 Information Resources Management Association International Conference*. Anchorage, Alaska, USA.

Straub, D., & DelGiudice, M. (2012). Use (Editor's Comments). *MIS Quarterly, 36*(4), iii–viii.

Treverton, G. F., Thomson, J. A., & Leone, R. (2001). *Reshaping National Intelligence in an age of information*. Santa Monica, CA: RAND.

Tropotei, T. O. (2018). Criticism against the intelligence cycle. *AFASES 2018, Scientific Research and Education in the Air Force*. https://doi.org/10.19062/2247-3173.2018.20.9.

Tyson, K. (2006). *The complete guide to competitive intelligence*. Chicago, IL: Leading Edge Publications.

Vuori, V. (2006). Methods of defining business information needs. In *Frontiers of e-Business research conference*. Tampere, Finland: Tampere University of Technology.

Wilson, T. (2000). Human information behavior. *Informing Science, 3*(2), 49–56.

Chapter 4
Business Intelligence Dimensions

4.1 The Notion of BI Dimensionality

The wide nature of insights expected from BI implies a search of a best-fit BI solution for a given organization. The existing variety of BI options creates a multidimensionality of choices (Skyrius & Nemitko, 2018) that encompasses both the information space (information sources, accessibility, efforts, costs) and BI functional space (methods, models, procedures, software products). This multidimensionality extends from basic figures and facts to meaning-rich contexts, from monitoring of everyday operations to the analysis of large one-of-a-kind problems (Davenport, 2015; Kamal, 2012), from internal efficiency to wide-ranging strategic insights (Dykes, 2016). The provision of multiple dimensions and heterogeneous views serves a more complete understanding about activities, opportunities and risks, and valuable insights may emerge as a result. Such understanding may require a certain instance of a BI system that is intended to respond to the variety of business issues and unified need for insights:

- Business is complex, dynamic and multidimensional, and the **variety of business issues** creates a task for BI to support the understanding of this complexity and detection of important issues. Context and complete view are of prime importance.
- The variety of business issues determines the **complexity and variety of information needs** for producing meaningful insights that are as complete and consistent as possible.
- The task of business intelligence to cover information needs defines its scope and structure as rather complex and multidimensional, requiring the **variety of tools and techniques**.

A business can be local or global, large or small, innovative or conservative, financially sound or ailing—all these features reflect some business dimension that can be used to define the business or benchmark against other businesses. The

R. Skyrius, *Business Intelligence*, Progress in IS,
https://doi.org/10.1007/978-3-030-67032-0_4

business intelligence system also has its degrees of freedom—it can be inward- or outward-oriented, centralized or decentralized, have wide or narrow coverage etc. BI cannot be everything to everybody in terms of intelligence needs, and the types of BI systems significantly differ even in the same industries. Existing cases of BI implementation show that there are certain practical BI dimensions alongside which a given BI system is implemented. The goal of this chapter is to discuss the key dimensions in business and BI implementation, looking for potentially important relationships that could prove useful for the success of BI implementation.

4.2 Business Dimensions

A given business situation that requires reliable insights or an important decision to make will have its set of the most important aspects or dimensions that, if properly evaluated, provide the basis for a fast and reliable estimate to act upon, as well as general awareness about the organization's standing and the state of its environment. Some of these dimensions define the own potential of the organization—e.g., financial strength or team and people quality. Other dimensions define external conditions on micro level (market situation and principal competitors) or macro level (legislation). A thorough evaluation of company's position on each of such dimensions would need an insight for whose production one or few simple procedures most likely would not be sufficient.

In the context of BI, the notion of business dimensions is sometimes used differently—OLAP technology, a tried-and-tested BI tool, assumes multidimensional analysis of data that reside in a data warehouse or an operational database. Here, multiple dimensions like time, location, product, buyer and others are used to keep an eye on anomalies and other issues of concern like growing problems or emerging opportunities. However, dimensions used in OLAP environment usually define a transaction or a group of transactions on a detail level. OLAP queries, although considered being above average complexity, produce exact answers from a detailed data in the source, and can be used as building bricks in developing an insight on some aggregated business dimension like the status of customer relations.

Examples of business dimensions have been provided in several sources (Dresner, 2017; Skyrius & Nemitko, 2018; Vuori, 2007) that have indicated a set of important features serving as business dimensions defining the type of business, its current environment and positioning along these dimensions.

Several examples of business dimensions from consulting, industry or professional sources are presented in Table 4.1:

Discussing BI adoption in Finland's construction industry, Vuori (2007) has pointed our several important dimensions that form the framework for a specific BI implementation in a given business organization:

Table 4.1 Examples of business dimensions

ProductiveFlourishing (Gilkey, 2011)	KPMG Industry 4.0 (Harris, Hendricks, Logan, & Juras, 2018)	Growth Decisions (The 8 Key Dimensions, 2015)	HTC consulting (Business Health Check, n.d.)
Strategy Operations Marketing Finances	Strategy Technology & systems Governance & risk management People Operational excellence Customer experience	Strategy Customer centricity Competitive context Organization and structure Culture Team management Management processes Support systems	Strategic Operational HR Marketing Customer Financial

1. Company level dimensions:

   (a) Structure and size—independent or part of a larger company
   (b) Resources—how much people, time and money are allocated for BI
   (c) Age, organizational culture and managerial attitudes—older companies might have a conservative org. culture that might affect managerial attitudes
   (d) Products—life cycle, sophistication, expertise-absorptive, length of value chain including subcontractors etc.
   (e) Markets—geographical scope, customer variety, market share
   (f) Competitive strategy—leader or follower

2. Industry level dimensions:

   (a) Clock speed—rate of changes and required reaction time
   (b) Competitive situation—number of competitors, entry ease or barriers
   (c) Culture, traditions and nature—traditional and sluggish or modern and fast

For example, construction business has a low clock speed and therefore hardly needs expensive real-time technologies to make intelligence information delivery faster, while yield management systems in travel business act on fast-changing demand data and use real-time approaches. As the aforementioned dimensions are expected to reveal key features and metrics, they form the basis for key performance indicators, or KPI, that satisfy the most important current information needs.

In 2008, a survey of small and medium business managers by the author (Skyrius, 2008) attempted to round up the most important business dimensions through important information needs. Data in the Table 4.2 from that research on the most important business information needs relate them to the corresponding business dimensions.

The points in the above table have been counted using the rating scale where for each respondent the most important type of information has received seven points,

Table 4.2 The most important business information needs: (Skyrius, 2008)

Information need	Points
Current status information—cash flow, liquidity, inventory, payables/receivables etc.	1429
Market information—dynamics, competition, innovations, trends	1188
Own performance information—how efficient is the creation of value; ratio of outputs to inputs	1079
Competence information—principal drivers of competence and competitive advantage, and their status	1048
Assets availability and use—what assets are used and how (in the first place—inappropriate use)	917
Macroeconomic information—figures and trends: inflation, interest rates, currency rates, prices of raw materials etc.	786
Other information—news, legal acts, etc.	24

second most important—six points, and so on—up to the least rated type receiving one point. According to the presented data, top points are given for the awareness of current business situation, and the more inert aspects regarding industry as a whole deserve less attention.

Based on the above arguments, the extended list of business dimensions that may be potentially important for BI system configuration and activities may be as follows:

1. Industry, its culture
2. Industry competitiveness—number of competitors, entry barriers; opportunity field—many or few;
3. Structure and size—independent or part of a larger company; large or small; local or global; market share;
4. Business age, organizational culture and managerial attitudes;
5. Inward- or outward-oriented;
6. Business uniform or diverse;
7. Clock rate—fast business with many transactions or slow business with few transactions;
8. Financial strength, profitability, asset use—good or weak;
9. Client relations—profound or superficial;
10. Team and people quality—strongly or weakly motivated;
11. Innovativeness—innovator versus follower; etc.;
12. Products—life cycle, sophistication, expertise-absorptive, complexity of value chain including subcontractors etc.

Some of the above dimensions will define the industry in which business operates "because business is done this way" (1–5), and some features of the industry like culture or clock rate will definitely have influence on BI activities. Other dimensions will define information architecture and KPIs to ensure proper monitoring (6–12).

4.3   Introduction to BI Dimensions

An assumption is made that the possible BI instances are, at least in part, driven by the need to handle the most important business dimensions. Other aspects of BI implementation may be driven by requirements of a certain BI technology, information behaviors and practices, organizational and intelligence culture. Some BI dimensions discussed further do not follow business dimensions, e.g., narrow or wide coverage; simple or complex questions; data-driven or user-driven; data detailed or aggregated—such dimensions are experienced in majority if not all businesses, regardless of industry.

BI dimensions are understood as important BI features that are most often mentioned in academic and professional sources when discussing BI implementation and adoption issues. The internal and external BI orientation has been discussed in (Cohen & Levinthal, 1990; Elbashir, Collier, & Davern, 2008). A scope of BI centralization from centralized to fully federated has been presented in a consultancy research paper by Accenture (Hernandez, Berkey, & Bhattacharya, 2013). A paper by Aberdeen Group (Krensky, 2015) analyzed empirical data on BI use and satisfaction, relating it to data silo and cultures of data ownership, and showing that data silos impede analytics and decision making. It has been shown that a growing BI system may migrate along the centralization-decentralization axis (Eckerson, 2013). The growing coverage of BI, or iterative development approach creating BI "islands", has been discussed in (Eckerson, 2013; Yeoh & Coronios, 2010). Among other published issues related to BI dimensions we can mention: real time to real value (Hackathorn, 2004; White, 2007); data-driven vs. needs-driven (Stangarone, 2015); system push or user pull (Lynch & Gregor, 2003). Such dimensions are important when available resources are limiting the scope of a BI system, and the positioning of BI functions along these dimensions to provide the most coverage and flexibility depends on the nature of business activities. In any case, the system is expected to provide a single consistent view of business environment.

A set of BI dimensions has been proposed in (Skyrius, 2015):

- Internal and external focus;
- Centralized and decentralized placement;
- Wide coverage versus narrow;
- Data driven (system push) versus insight-driven (user pull);
- Real time versus right time.

Timo Elliott (2004) has proposed one more set of BI dimensions:

1. **Data depth**. Not all users need the same level of information detail. For example,

executives might want highly aggregated information, while front-line employees might need detailed order data. Still other users may need access to real-time data feeds, directly from the operational systems. Because these different types of information are often stored in different systems and formats, this may affect product choices.

2. **Data breadth.** Some users may need to compare information across different systems. It is unlikely that all the information that users want to access is stored in a single data warehouse. The ability of the BI tool to do cross-functional analysis may be a determining factor.

3. **User control.** While some users need to access information directly and autonomously, others may need specially preconfigured content. A chosen BI standard should ideally support both user types.

4. **Ease of use.** It is important that the standard BI tool enable smooth transition from basic to more sophisticated interfaces.

5. **Customization.** Some users will need more customized interfaces than others. Executives may need dashboard interfaces, or embedded reports in custom applications. The ideal BI standard provides a single global framework for BI deployment, whether users access data through standard tools or customized interfaces.

6. **Business specialization**. Some users may need content or functionality that is specifically designed for a particular industry or business domain. For example, a certain BI standard may have products specifically designed for human resource managers or finance staff.

On the basis of proposed set of BI dimensions, the joint list of BI dimensions can be suggested:

1. Internal and external focus.
2. Centralized and decentralized deployment.
3. Question complexity—simple to complex.
4. Functional scope—wide to narrow.
5. Automation—manual to automated.
6. Initiative—data driven (system push) versus insight-driven (user pull).
7. Real time versus right time.

4.4   Relations of Business and BI Dimensions

The reliable assessment of business dimensions is well argued in (Ramakrishnan, Jones, & Sidorova, 2012) and is targeted towards producing **a single consistent version of business information**. In the informing process executed by BI, the key business dimensions need to have reference points, often in the form of key performance indicators, or KPI. The role of such indicators is to unify the most important business metrics. The benefits of such unification are obvious: better data quality; more reliable arguments for decision making; facilitation of communication between stakeholders. For these dimensions to be seen and evaluated together as a whole, many technological issues have to be solved, like information integration from different sources, integrated presentation and visualization, flexible software for dimension manipulation, to name a few. Such issues lead to BI implementation options and technology dimensions that are expected to support the required flexibility along business dimensions.

The projection of business dimensions into BI options would lead to requirements for important BI dimensions and operating modes, e.g.:

1. What data is used or specifically collected for BI purposes.
2. Is it collected/located centrally or locally.
3. Is the data more internal or external.
4. Functional scope—narrow "islands" of intelligence or analytical functions, or company-wide views across functions.
5. How fast intelligence/analytics have to be performed (latency).
6. System push or user pull? (data-driven or insight-driven?).
7. Automated or manual, or automated to what extent (how many repeating procedures).
8. Complex or simple questions (simple or advanced approaches and tools).
9. How detailed the data have to be.
10. How measurable, computable and IS-covered the data are.
11. Monitoring permanent or punctuated.

The following Table 4.3 is a projection of business dimensions into BI dimensions, based on how do information needs for a certain dimension/activity translate into BI dimensions/functionality.

The following are several examples of projecting business dimensions into BI dimensions, showing how information needs for a certain dimension or activity translate into BI dimensions and functionality.

Financial strength is monitored internally; the function is centralized; has wide coverage for entire business but limited BI tool palette; questions are simple but need encompassing and reliable assessment; because of strict and stable rules much of insight production can be automated; significant and potentially automated system push and alerts coupled to user pull when analysis and flexibility (multidimensional, what-if) are required; right time information delivery to be able to make right and timely decisions (right is more important than timely).

Team and people quality is monitored mostly from internal sources except talent search and hiring; the function is centralized; wide coverage but simple tools; questions or issues are simple; automation potential for simple functions like tracking of performance; information delivery is activated mostly by user pull; right time delivery—no need for fast reaction.

Client relations—monitored both internally and externally; the function is centralized and customer-centric; a process loop joining operational and analytical parts with feedback; data integration is essential; velocity depends upon business speed; significant role of analytics (one of the most analytics-consuming activities); IT coverage of raw primary data should be as encompassing as possible.

Table 4.3 A projection of business dimensions into BI dimensions

Business dimensions	A. Internal and external focus	B. Centralized and decentralized placement	C. Wide coverage versus narrow	D. Question simplicity vs. complexity	E. Automation	D. Data/insight driven	E. Real/right time
Structure and size—independent or part of a larger company	Internal						
Age, organizational culture and managerial attitudes	Internal						
Financial strength	Internal	Centralized	Wide (entire company)	Simple	Partial	System push	Right
Team and people quality	Internal	Centralized	Wide	Simple	None	Both ways	Right
Client relations	External; related to internal (clients, sales)	Both, depending on the nature of business[a]	Narrow but important and not isolated	Simple to complex[b]	Partial	Both ways	Both, driven by business speed
Markets and opportunity field	External	Decentralized, but possibly federated to put the pieces together[a]	Wide. Integration of rather heterogeneous information; composite meanings and insights are very important but take significant effort to achieve; "Black swan" syndrome in detecting unusual cues	Simple to medium[b]	Partial	Both ways	Right time

Competition and key competitors	External		Wide. Integration of rather heterogeneous information (especially competitive intelligence)	Simple to medium[b]	Partial	Both ways	Right time
Competitive strategy and strength; innovativeness	Internal and external to benchmark	Centralized	Wide	Simple to medium	None	Insight-driven	Right time
Products	External	Decentralized, mostly in R&D	Narrow	Simple to medium	None	Insight-driven detecting innovations	Right time
Asset use: relaxed to strained	Internal	Centralized	Wide	Simple	Partial to full	Data-driven (acceptable limits, alerts)	Right time
Industry clock speed[c]	External						
Industry competitiveness[c]	External						
Industry culture[c]	External						

Source: author

[a]Hard to centralize because of the multiple sources of incoming information; federation/integration required to attempt a more complete picture

[b]Client data from CRM-like systems may undergo rigorous and complex analysis because of its structured and organized nature; they may be deemed analysis-friendly. Market and competition data is less prone to complex analytics because of its variety and complicated integration. As well, its provenance and reliability is far more questionable

[c]The specified industry features are slow-changing and therefore do not require permanent monitoring; they might be at the beginning of the list because their instances largely set the intelligence mode—different intelligence approaches will be used for fast or slow industries, cutthroat competitive or calm and settled, conservative or innovative culture etc

4.5   BI Dimensions

4.5.1   Internal and External Orientation

An often-mentioned feature of BI is its separation into internal and external intelligence. The latter is often termed "competitive intelligence", although this term focuses mostly on competition and does not cover the other important elements of external environment—customers, suppliers, partners, innovations, legal environment etc. There are significant differences in producing internal and external intelligence (Skyrius, 2015):

- **Internal intelligence** is based on strong signals, hard facts and clarity; performs almost entirely automatic data integration by hard rules; receives its feed mostly from internal information systems whose behavior, product and logic are known and under control;
- **External intelligence** seeks to make sense of many weak signals from external environment; the true value of data is often unclear at its reception, and revealed only by experimenting or after it has been used; information integration is executed by soft rules, middleware, passive integration; information handling processes are rather chaotic with mixed flexibility and provenance.

Tyson (2006) defined competitive intelligence as a systematic process to convert fragments of competition information into strategic knowledge for decision making: current competitors' position, historical performance, strengths and weaknesses as well as specific future intentions. However, intelligence objectives presented in (Tyson, 2006) are equally applicable both to internal and external environments:

- To avoid surprises;
- To identify threats and opportunities;
- To gain competitive advantage by decreasing reaction time;
- Improve long and short term planning.

This supports the earlier conclusion that the division of business intelligence and competitive intelligence into their respective areas of internal and external interest contradicts with the very meaning of the term "intelligence". Intelligence presumes the level of understanding that is comprehensive and supported by possession of varied, multidimensional and mutually supportive information.

In previous research (Skyrius & Bujauskas, 2011) the author has examined the results of a survey on business information users, concentrating on the complex side of their information needs. The questionnaire was distributed among several groups of respondents mostly employed as middle-to-senior business managers; the surveyed entity could be regarded as a convenience sample in the community of business decision makers, totaling in 203 responses. The part of the research dealing with features of complex information needs had placed external information among top three features specific to complex information needs—107 cases out of 203. Several issues in the above research have revealed combined parallel monitoring of

Table 4.4 Groups of most frequently monitored information (Skyrius & Bujauskas, 2011)

Information group	No of cases	Type
Competition and market information	136	External
Accounting and finance	100	Internal
Customers	72	External
Sales and turnover	62	Internal

both internal and external information. The four top groups of information monitored most frequently are (Table 4.4):

As we can see from the data in Table 4.4, the most frequently monitored data comes from both inside and outside sources. Many business questions can cross organization boundaries, like the examples below (Skyrius & Bujauskas, 2011):

- The dip in sales that might be related to poor promotion efforts, substandard logistics, dubious production quality control, or faulty components;
- Weakened cash flow, forcing to take a closer look at income and costs, both related to outside subjects (customers, suppliers and partners), as well as inside factors—inadequate cost control, too much manual labor etc.;
- Unexpected spike in received sale orders that can be attributed to anything from internal factors (e.g., extra functionality of a product or its ability to connect and operate easily with related products) to external ones (e.g., interest from unexpected kind of buyers, or stumbling upon a previously unknown market niche).
- Rumors of a major merger in the industry that would significantly change the market's competitive structure; although being external in its nature, this threat requires to mobilize resources and prepare best possible countermeasures.

In dealing with such questions, the sources of information cannot be purely internal or external. The decision making processes have to evaluate a sufficient set of internal and external factors, leading to a best possible understanding of a problem situation and its outcomes. Cohen and Levinthal (Cohen & Levinthal, 1990: 133), discussing absorptive capacity, wrote: "With regard to the absorptive capacity of the firm as a whole, there may, however, be a trade-off in the efficiency of internal communication against the ability of the subunit to assimilate and exploit information originating from other subunits or the environment. This can be seen as a trade-off between inward-looking versus outward-looking absorptive capacities. While both of these components are necessary for effective organizational learning, excessive dominance by one or the other will be dysfunctional." Meredith, Remington, O'Donnell, and Sharma (2012) have related internal or external BI orientation to BI development approach, and noted that early increments of successful BI systems often start small and have a narrow scope (for example, a departmental data mart using data from one or two systems). This early increment of the BI system can only enhance the inside-out capability, centering on internal functions. Later increments will increase the scope of business functions and enhance the outside-in (internal) capability, assuming the scope of the system grows to include more internal systems and, eventually, external systems and data sources. This point has been supported by

Miller (2000): "It is not possible to understand a competitor unless you understand your own company first".

The required coverage of business environment has to be supported by appropriate mix of internal and external intelligence, providing a richness of context that adequately serves the activity needs. The analysis of this mix might provide insight into situations where seemingly unrelated cues, based on previous experience, can point to a growing problem or an opening opportunity. Once this mix of information had justified itself, users can be aware and take a proactive stand. For a given organization, the exact mix of internal and external orientation will depend on many factors like the dynamics of competition or key business priorities.

4.5.2 Centralized and Decentralized BI

Centralization and decentralization should reflect the nature of business—be it rather uniform or diverse, centralized or distributed. Accordingly, BI resources and control can be centralized, decentralized or exist as a blend of both. An example could be General Electric corporation that at one point have owned a TV broadcaster, a jet engine division, and a finance division; all these diverse business lines could hardly require a single centralized model. Although organization-wide BI systems are intended to provide the necessary transparency, there are numerous opinions on their inertia and in favor of self-service BI (Eckerson, 2013; Krensky, 2015). Back in 1990, Cohen and Levinthal (1990) had stated: "When information flows are somewhat random and it is not clear where in the firm or subunit a piece of outside knowledge is best applied, a centralized gatekeeper may not provide an effective link to the environment. Under such circumstances, it is best for the organization to expose a fairly broad range of prospective 'receptors' to the environment".

On the other hand, fully decentralized BI might scatter intelligence resources among "intelligence islands" or shadow systems; a proof to this is the continuing use of spreadsheets in analytical functions. The benefits of self-service—more flexibility and speed, faster adoption, and better alignment with business needs—have to be balanced with protection against potential problems: duplication of data and tools and waste of resources, lack of business-wide insights, and unclear responsibility.

Both approaches have their strong and weak points; however, strongly centralized BI systems seem to be favored less because of their size, inertia and inflexibility. Centralized BI approaches are advocated by big players in BI market—IBM, Microsoft, Oracle, SAP, SAS, who offer large-scale architectures and databases, centralized management and control. However, a centralized BI model is slow to adapt and costly to implement. According to Meredith et al. (2012), "large-scale, inflexible, slow-to-adapt BI systems cannot provide the kind of decision support needed in a fast-paced changing market". A purely centralized BI system is a rare case if in existence at all. Many published cases on company-wide BI solutions (e.g., Eckerson, 2013; Elliott, 2004; Krensky, 2015) stress the necessity to achieve a balanced mix of centrally managed standards on one side, and flexibility for business user special needs on another. For a rational balance, there are things that have to be centralized for the backbone of the solution to run reliably: a single source of truth; standard procedures for accessing and using applications; units of measure; standards for organization-wide processes such as inventory management, quality management, human resources, research and development; shared analyses and insights, to name a few. An important issue here is a sensible set of common standards applied throughout the organization, where too few standards create challenges to integration, and too many standards impair flexibility.

Company-wide BI is not necessarily a strictly centralized system, or a strictly self-service system, though proliferation of easily usable BI tools makes it easier for casual users to master the BI skills. The most effective BI units, as defined by Liautaud and Hammond (2001), are the ones that apply "information democracy", rather that the extremes of "information dictatorship" and "information anarchy". BI will not achieve a required scope when it is isolated, so the mix of centralized-distributed BI should be based on a rational balance point, leveraging the strong points of both approaches. An approach suggested by Ralph Kimball for building data warehouses supports the development of BI system from separate part into a whole: the process can be of "bottom-up" type, starting with manageable data marts centered around business processes and reaching across functional units (Ross, 2001). Specialist sources (Standen, 2010) point out that in a distributed structure there might be a diversity of decisions. This diversity may lead to both right and wrong decisions, but it is expected to bring out innovative approaches, resulting in insights that might be neglected or avoided in a centralized case. With distributed structures, users get more freedom in their information activities, building and rebuilding their intelligence models as the environment changes. To prevent fragmentation and duplication of efforts, exchange of information and expertise between users is necessary. As the same source (Standen, 2010) states, "The existence of shadow systems, and the extent of them, is the clearest argument that centralized business intelligence alone is simply not up to the task". To the author's opinion, this statement, reinforced by the benefits of distributed approach, illustrates the need for an intelligence culture based on horizontal propagation of information. This point has been supported by Tyson (Tyson, 2006), who had stressed that before any kind of IT-based supporting system is implemented, it is necessary to develop an intelligence culture supporting sharing of intelligence information.

A selection between centralized and decentralized approaches inevitably has to deal with decisions on the set of standards for BI activities—coverage, data governance, security etc. Boyer, Frank, Green, Harris, and Van De Vanter (2010), citing a white paper by Ventana Research, provide several key points to consider when supporting decisions on standards:

- **Broad coverage**—a standard should encompass as much of the organization as possible; cover key business processes; and ensure compliance with regulations;
- **Flexible architecture and heterogeneous data access**—use of well-known IT standards like XML, SOAP or WSDL to integrate disparate environments, thus

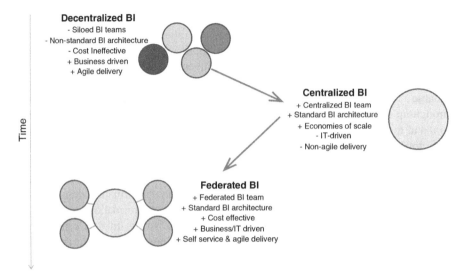

Fig. 4.1 Evolution of BI organization (Eckerson, 2013)

reducing the number of required iterations for producing intelligence in the future;

- **Scalability**—flexible server technology to support load fluctuations and future growth;
- **Global capability**—in the case of multinational or global activities, capabilities should be accordingly compliant, multi-lingual and multi-currency-capable;
- **Security**—a BI standard should leverage the existing enterprise security.

Eckerson (2013) proposed federated BI as a solution combining the strong features of both approaches: agility of a decentralized approach and consistency of a centralized approach (Fig. 4.1).

According to Eckerson, a federated BI approach maintains a unified BI architecture that delivers consistent, accurate, timely, and relevant data. At the same time, due to locally placed BI professionals, business analysts finally become proficient with self-service BI tools. In addition to this, BI solutions are delivered in much less time, improving BI agility under ever-changing conditions. Of course, challenges remain like managing matrix relationships and balance between business and IT.

It is rather obvious that the balance of centralized and decentralized approaches will differ from organization to organization, and a tradeoff in the form of competence center is one of the ways to support this balance. The emergence of business intelligence competence centers (BICC) around the beginning of the new millennium has been targeting exactly these challenges. In a consultancy research paper by Accenture (Hernandez et al., 2013), an evaluation of BI and analytics function placement states the importance of a dedicated organizational unit—BICC or Center of Excellence—to support the blend of control and flexibility. BI evangelist Timo

Elliott (2014) supports the establishment of BICC as a cross-functional vehicle to support BI and analytical functions across an organization.

To be exact, both sources quoted above talk about analytics, not BI; however, as discussed earlier in Chap. 3, BI and analytics do not differ much in their goals and functionality. Use of the term "Business intelligence Competence Center" to define placement for an analytics function in quoted sources only justifies this absence of difference.

4.5.3   Question Complexity

On the dimension of question complexity, the instances range from **simple** informing questions requiring few sources, procedures and functional tools to **complex** or complicated informing questions, requiring integration of information from a number of sources, a wide set of support procedures and tools. This variation in complexity requires different approaches in terms of insight development process and tools.

The rate of problem complexity relates to a framework proposed by Snowden and Boone (Snowden & Boone, 2007), which defines four types of problem domains:

- **Known** (not to individuals, but to organization): clear relationships, repeatability; actions—sense, categorize, respond;
- **Knowable**: can move to known, but lacks understanding and resources to learn; actions—sense, analyze, respond;
- **Complex**: large number of agents and relationships, patterns are perceived but not predicted, repetition is random: action—probe, sense, respond;
- **Chaos**: no perceivable relations between agents, or between cause and effect; turbulent environment; action—act, sense, respond.

Complex domains differ from chaotic domains: in complex case we can look for patterns, when in chaotic case we need to stabilize the situation in order for patterns to emerge. Snowden and Boone did not specifically relate their framework to information needs and possible subsequent forms of IT-based support, but we can infer that as the uncertainty of problem domains increases, IT support gets more complicated, and the proposed classification scheme can be useful defining the features of complex environments for subsequent selection of most appropriate approach.

The analysis of various dimensions of information needs shows that growing uncertainty and complexity reduce the potential of automation of informing functions, at the same time raising requirements for their flexibility and variety. The approaches for satisfying complex, problem-specific and non-repetitive information needs are hard to "softwarize", and therefore present a specific field for research interests.

4.5.4 Functional Scope

The functional capacity of BI techniques ranges from simple reporting and graphing to advanced and flexible features: performance management, predictive and prescriptive analytics, data mining, text mining, data and information integration, machine learning analytics, and others. These features are expected to be available across different platforms, including mobile and cloud-based deployments. The sophistication level of advanced BI instruments often brings up a need for a user to master a complex task model as required by the instrument, adding to the complexity of a problem at hand.

There might be different cases requiring different sets and scope of BI techniques:

- Wide array of tools—a palette of assorted BI techniques for all domains and stages of BI process; likely for a complex and diverse set of activities;
- Few tools—situation-specific or activity-specific tools, techniques, models; likely for one or few key business aspects.

The data from earlier research, mostly on business decision support and IT role in providing support (Skyrius & Winer, 2000), has shown that in many intelligence and decision support cases the structure of functional support splits into must-have and just-in-case groups of functions. Two groups of respondents, one from Lithuania and another from USA, represented small and medium businesses and were asked to indicate what processing modes they apply to the source information for decision development. The results are given in the Table 4.5 and in Fig. 4.2.

It is interesting to note that, despite the statistically small data volume, the ranking of the processing modes is identical, the priority being given to the simplest modes that, if successful, are the fastest and least expensive ways to required insights. A tiered approach might serve as a solution in developing an efficient functional space, where the first (close) tier would contain the most often-used tools, and the more complex functions that have a less frequent use would be moved to second tier. A more detailed discussion on tiered approaches can be found at the end of this chapter.

Table 4.5 Use of processing modes for decision information (Skyrius & Winer, 2000)

Processing mode	Cases in LT	Cases in US
Browsing	34	14
"What-if" analysis	25	13
Sorting and grouping	20	11
Graphing	14	11
Pattern tracking	10	9
Mathematical analysis	7	7

Distribution of processing modes

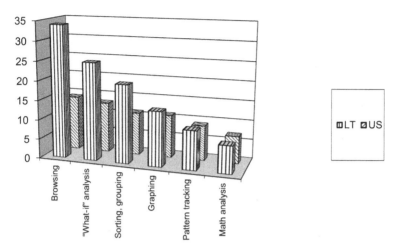

Fig. 4.2 Distribution of processing modes (Skyrius & Winer, 2000)

4.5.5 Automation

The growing uncertainty and complexity of information needs reduce the potential for automation of informing functions, at the same time raising requirements for their flexibility and variety. The potential of "softwarization" applies best for routine repetitive procedures, while it is hard to automate the procedures for meeting complex, problem-specific and non-repetitive needs. One of the prospective avenues raise effectiveness of complex informing is to seek the best possible balance between user functions and IT functions. Quite a few research publications (e.g., Abbasi, Albrecht, Vance, & Hansen, 2012; Berkman, 2001; Carlo, Lyytinen, & Boland, 2012; Straub & DelGiudice, 2012) have stated that IT has to support users in a way that boosts their intelligent powers, instead of aiming to substitute them.

As information tasks differ by their routine or non-routine nature, we can roughly fit them into three groups (Skyrius, 2014) by their automation potential:

1. Examples of tasks with high level of automation:

   - Database querying,
   - Web search,
   - Arranging and combining data,
   - Recommenders.

2. Examples of human tasks mostly based on heuristics and hard to automate (low

potential of automation):

- Semantic integration,
- Sense-making,
- Causal analysis,
- Creation of alternatives,
- Negotiations.

3. Examples of "gray area", where sophisticated support and synergy between IT and user are sought (medium level of automation):

- Integration of data—manual or automatic;
- Influence modelling;
- Rule creation and use;
- Text analytics;
- Simulation, "what-if" modeling.

In terms of possible synergy between IT and users, the third group is the most ill-defined. The definition of this "gray area" is complicated by the existence of some vague lines between automation and human action: for instance, we can fully automate a simulation experiment, or we can leave all emerging "what-if" questions to be handled by human actors. Similarly, customer behavior rules can be induced by the data mining software, or they can be inferred by analysts interpreting the results of analysis. Currently this group is attracting rising interest from developers of AI applications.

Same automation issues come up when dealing with information integration as one of the key functions in BI. We may perceive intelligence as largely a science and art of sense-making by putting information together. If we attempt automating information integration, we attempt to automate sense-making. The features of information integration therefore are close to those of BI in general—complexity of function (simple integration, complex integration); expectations from automation and existing limitations. Integration approaches differ for structured and unstructured information. For the former, integration is mostly based on relational instruments from the database field, and for the latter the range of techniques is much more diverse, starting from manual integration and going all the way to natural language processing (NLP) and other advanced techniques. The issues of information integration are discussed in more detail in Chap. 5 of this book.

Several examples of attempts to automate the fulfillment of complex information needs are a set of technologies called Complex Event Processing, or CEP (Hasan, O'Riain, & Curry, 2012; Luckham & Frasca, 1998), Meaning-Based Computing, or MBC (Huwe, 2011), optimized real-time business intelligence (Walker, 2009). In the case of CEP, to the author's opinion, several important problems come up that might impair its adoption:

- The combinations of signals are related to the business rules or knowledge structures in expert systems. Known combinations are based on available experience; unknown combinations are hard to evaluate, and a "black swan" problem arises—how to recognize new important phenomena.
- Does the performance of a system intended to recognize important signal combinations depend on the repetition ratio of the combinations?
- How the multiple and often overlapping signal combinations are going to be interpreted? Is it the task of people or software?
- If the number of events and signals is rather high, on what principles the volume of information presented to the users will be reduced? Is it going to be grouped, aggregated, or distilled into some derived sense?
- Similar to Big Data problems can be expected: a huge number of signals might produce many derivative combinations, whose meanings, importance or reliability are rather hard to verify.

While discussing the issues of IT-driven automated efficiency versus user-handled flexibility, one more issue emerges. It is possible that automatic informing can actually *strip the context* of informing by being too exact to the request. If the rules that do the delivery are trustable and conveniently reduce the information overload, this approach can potentially create an information *underload*—a context-poor environment that makes users question the results of automatic informing, all the more if the logic behind the result is complex and opaque. For example, a complicated yet direct question (e.g., "What is our market share?") might drive the system to provide a singular and correct answer (e.g., 16.7%). Yet, as context research shows (Hasan et al., 2012), this answer without appropriate context might raise doubts about its validity. This can lead to an idea that in satisfying complex information needs the effect of the "last mile" of informing activities left to users would maintain the richness of context.

4.5.6 Initiative: Supply-Driven Versus Demand-Driven

The dimension of initiative is seen here as a range between demand-driven (or question-driven) and supply-driven intelligence. In the demand-driven case there is a question, an assumption or a need to develop an insight, which in its own term creates an information need, or demand. In the supply-driven case there is an available resource—data, software or models, and their existence is expected to create value by using them to produce information which is assumed to be valuable.

Recent developments in the field of BI, especially dealing with Big Data, have produced some new terms, including a term of "data-driven". Here, a confusion of terms is possible: supply-driven might be confused with data-driven, although the exact meaning of the latter is still unfocused. Kasibhatla (2017) talks about data-driven culture, while at the same time referring to insight-driven organization that needs an appropriate culture to be built. Dwivedi (2016), while discussing a data-driven approach, actually talks about *information* required to back up decisions, not data—raw, unprocessed, context-less data seldom may serve as an argument in decision making.

Various sources define data-driven approach to intelligence and decision-making as the one based on data availability and analytics, as opposed to gut-feeling and common sense without many hard arguments. However, pure gut feeling-based decisions are hardly imaginable in contemporary business—since early applications, information systems meant exactly reducing uncertainty and ambiguity in managing business. This is rather true regarding the more advanced decision support systems whose prime role always has been a reduction of uncertainty by providing solid, fact-based decision arguments.

With initiative, there is a similarity to another dimension that has been in use for some time regarding the split of initiative between the system and the user: *system push* and *user pull*. Supply-driven BI initiates system push; its tools are mostly pre-built models, alerts, rules and applications. Demand-driven or question-driven BI works on user pull and often is reactive, although there may be insightful questions directed into the future and at the same time based on the understanding of present and past.

Supply-driven approach has a lengthy history of generating reports with aggregate relevant information in order to support management processes (Bucher, Gericke, & Sigg, 2009), which rather reminds the mission of MIS. On the other hand, alerts and other cases of automated informing in proactive systems belong to a supply-driven approach as well; there are numerous examples of supply-driven intelligent functions: complex alerts, flexible inventory management, adaptive scheduling and others. These functions are of regular nature and can form a basis or a part of monitoring system; this should mostly apply to regular data from operations regarding internal environment and close external environment (customers, suppliers, partners, competition). When "radars light up", indicating an issue worth attention, a need emerges to move from passive to active stand to learn more about the issue of interest.

Demand-driven approach (or question-driven, or issue-driven) is based on user needs: a business problem or question starts the process to create required sense—test an assumption or investigate something that has attracted attention. Users look for the data sources that they need in order to address those problems, and then for cues and patterns in data that lead to the required sense. This approach well conforms to two tier support structure mentioned earlier in this chapter. The demand-driven approach has been widely used by decision support systems that had addressed certain problems as they emerged. Daly (2016), presenting a model of linking decision support information supply and demand, has specified several decision problem abstraction levels, where simple and least abstract problems allow structured approaches, and most abstract problems do not allow solution formulation because of their complex and unstructured nature. Daly also supported the point that the growing abstraction of a decision problem both diminishes formal capacities of a decision support technology, and raises the importance of people participating in the decision making process.

A well-directed business question is an important issue; quite often analysts have to clarify the business request with the source of that request, sometimes undergoing several iterations. A vague question has to be reformulated regarding the boundaries of a problem, principal components and context, decision criteria. The adequate clarification of these issues defines more precisely the resources required, procedures to be executed, and generally saves time and effort of the participants in the intelligence process.

A side issue of demand-driven BI is self-service BI and analytics, having gained prominence in recent years. The roots of this phenomena rest firmly in the use of spreadsheets, MS Excel in the first place; the existing so-called Excelization of BI and analytics is confirmed by numerous sources. While self-service BI encourages users to base decisions on data instead of intuition and removes the middlemen, it creates the risk of chaotic reporting within an organization. To reduce this risk, a basic set of standards like key metrics and a unified business vocabulary, mentioned earlier in this chapter, should be determined together with processes for creating intelligence reports and publishing them.

The ability or skills to formulate well-pointed business questions seems to be on the rise with data analytics community, and, to author's opinion, may become one of the key analytic competencies, together with statistical and tool-oriented skills. Hernandez et al (2013) state that "companies need to experiment with developing mechanisms that identify, track and realize the value of Analytics efforts." The value realization mechanism will help move organizations from a data-based mindset to an outcomes-based mindset and minimize needless or unproductive requests for analytics support.

4.5.7 Velocity: Real Time Versus Right Time

The notion of real-time business intelligence has been around for the last 10–15 years, implying fast production of decision information required in high velocity activities. Its opposite is the so-called right-time business intelligence, so this dimension would cover BI systems from systems delivering results in fastest time possible, or **real-time** systems, to systems using needs-defined latency, or **right-time** systems.

As different industries feature different clock speed, some fast-moving industries have more potential to benefit from real-time BI by enabling faster insights, decisions and actions. Although real-time BI technologies are expensive (Antova, Baldwin, Gu, & Wass, 2015), the expected payoffs outweigh the costs. Anderson-Lehman et al. (2004) have discussed the case of Continental airlines getting from worst to first in the industry, where a large share of success has been attributed to a number of key applications rely on real-time data: fare design, recovering lost reservations, marketing insight, flight management dashboard, fraud investigations, to name a few. There are other areas where high clock speed carries potential benefits from real-time BI. Some of those areas are (Table 4.6):

Table 4.6 Activity areas with high clock speed

Fast business	Algorithmic trading (or, more precisely, automated trading)—uses set of rules and complex models; may create issues because of absence of human risk controls and "soft" evaluation Demand sensing—models of fast-moving inventory Dynamic pricing and yield management—travel business; reaction to demand changes Customer relationship management—quick analysis to react to issues coming up like customer inquiries, changes in orders, complaints etc. The goal is to have a fast and well-informed reaction, based on logic of rules and analysis of historical data. Challenges include information integration and interdependencies between channels.
Fast internal operations, monitoring of assets	Systems monitoring—real-time IS management tools; application performance monitoring Utilities monitoring—identification of potential problems in asset infrastructure Supply chain optimization—optimal placement of inventory; optimized parcel loading for reduced overall delivery times
Detection of risks and evil intent	Payments and cash monitoring—credit card fraud detection; detection of a stolen card Data security monitoring—real-time intrusion detection RFID/sensor network data analysis—early detection of disasters, calamities etc.

Source: author

An earlier survey, performed by the author (Skyrius & Bujauskas, 2011) on a convenience sample of 203 business students at masters level, has questioned the respondents to indicate up to five types of information being monitored most often in their activities, together with monitoring latency. The results of the survey have been processed using IBM SPSS Modeler v.14 data mining software, relating the type of monitored information to its latency. The graphic view of the relationship is presented in Fig. 4.3. Table 4.7 presents the actual numbers of links. The related elements are coded as follows:

1. **Information types:**

   S—sales;
   A—accounting;
   L—inventory, stock;
   R—market and competition;
   K—customers, customer relations;
   T—legal acts;
   P—projects;
   D—personnel, employees;
   G—production, business processes.

2. **Information latency:**

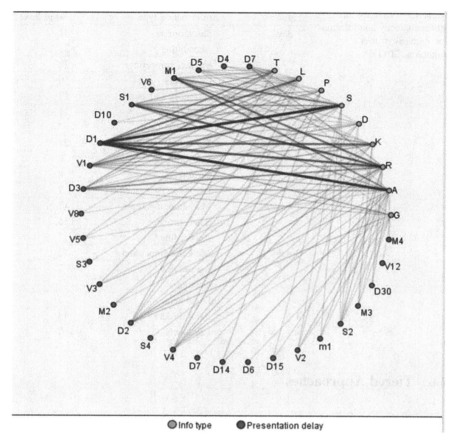

Fig. 4.3 Relationships between monitored information types and monitoring frequency (Skyrius & Bujauskas, 2011)

V—hours: V1—1 h, V8—8 h etc.;
D—days: D1—1 day; D3—3 days etc.;
S—weeks: S1—1 week; S2—2 weeks etc.;
M—months: M1—1 month; M2—2 months etc.;
m—years: m1—1 year.

The 1-day latency, as shown by the most frequent links, dominates with the related types of monitored information, among which both internal and external information types are important.

A convenience sample does not allow for any statistically sound conclusions; moreover, it did not include participants from "fast" businesses acting in capital, commodity or mobile communications markets, or employing real-time e-business models. However, the data from the surveyed entity show that the urgency of information presentation in most cases is defined by the real "speed of business".

Table 4.7 Strongest links between latency and information type (Skyrius & Bujauskas, 2011)

Latency	Information type	No of links
1 day	Sales, orders	31
	Accounting	31
	Market, competition	22
	Customers	19
	Inventory and assets	17
	Legal acts	11
	Processes	7
1 week	Market, competition	21
	Sales, orders	9
	Customers	8
	Legal acts	7
1 month	Market, competition	17
	Sales, orders	15
	Accounting	13
3 days	Market, competition	12
	Accounting	7
	Customers	7
	Legal acts	7
1 h	Sales, orders	14
	Legal acts	7

4.6   Tiered Approaches

In search for a compromise decision to select the configuration of a BI system that would be encompassing and at the same time flexible, a tiered approach, mentioned before in this chapter, might be a viable option.

A tiered approach relates to the process of building an insight or solving a problem by starting in the closest environment with available data sources and tools. The process steps are expected to reduce uncertainty in multiple possible ways: going from "bird's-eye view" to "worm's-eye view"; from aggregated data levels to finer detail; moving along some data dimension to compare different instances, e.g., "how did this look like a month or few months ago?" or "what is the situation in other branches?" The uncertainty under attention should be diminished, and if the tools of the first tier are not adequate, a need emerges to move to the second tier or further tiers, as shown in the Fig. 4.4 below.

The author has earlier suggested (Skyrius, 2014) to support the balance between opposites in different BI implementation dimensions by splitting the BI environment into two tiers:

- The first tier containing a simple set of support tools that are close and easy to use;
- The second tier containing more distant and more complicated information sources and processing techniques that are required much less often;

Fig. 4.4 Relation between uncertainty and tiers of informing environment

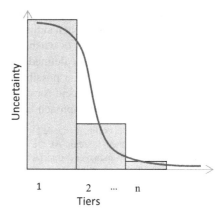

- Manageable support environment that allows easy switching of items between tiers, similar to the form of managerial dashboards with interchangeable items on display.

The **first tier** should contain the close support environment that is required most of the time, is simple in its use and can be easily configured:

- Basic data on internal and external environment: sales, market share, cash-at-hand, order or project portfolio; comparative figures by time, place, product etc.;
- Search and arranging tools: simple search in own sources (databases, archives, competence and experience bases); simple search in public sources—generic (e.g., Google) and specialized (e.g., Financial Times archive); tools for arranging search results (e.g., by relevance or size), easy classification and annotation;
- Tools for simple calculations—templates, simple models.

Second tier might, for example, include:

- More specialized and complex information sources with advanced search and information integration tools;
- Modeling tools for forecasting, scenarios, simulation;
- Data analysis and presentation technologies—drill-down tools, OLAP, data mining facilities; graphing and visualization tools.

The existence of tiers beyond the second tier is more of a theoretical nature—it may be, e.g., cloud-based specific powerful analytical technologies that are seldom used.

The features of the first tier offer simple functional support and quite closely match the technology of managerial dashboards, whereas the second or more tiers aim at specific problem solving and decision support needs, and can be used in specific cases requiring sophisticated tools and techniques. A need to "dig deeper" is expected to engage into more powerful analytics; often this is the case of switching from more general functions of BI to decision support tools. Such split of functionality would roughly reflect required functions for generic business intelligence and decision support cycles respectively. It would also allow for required cross-functionality in the cases when simple decision support needs would be well-served by first tier functions alone, or when business intelligence needs would require more advanced tools. The more defined set of features for both tiers of the support environment could lead to a possible set of requirements for the interface design of an information environment for decision makers.

An example of tiered approach is presented in (Mayer, 2010).

> The CIO at a global company needed to take an end-to-end view in designing his company's business intelligence system. At the front end, reports needed to cascade in a logical hierarchy. The top managers' interface required an intuitive design offering a wide range of analyses. To support this robust interface, the underlying information structure would have to be reconceived. Data were standardized and information pathways reworked for the most important key-performance indicators (KPIs) needed to manage the company. The corporate navigator is its framework for designing a next-generation executive information system with the four-layer business/IT model shown in the exhibit.
>
> The entry screen, which gives executives a graphical top-line summary of the most important KPIs (Layer 1), includes comment fields to highlight findings or explain deviations. Using the one-page reporting format, Layer 2 offers an at-a-glance assessment of corporate performance in four core areas: financial accounting, management accounting, compliance, and program management. With a few mouse clicks, anyone interested in the underlying details can access pre-defined analyses, such as contribution margins and P&L calculations, by product or business unit. A hierarchical reporting structure connects these views, allowing users to toggle through different levels of performance data.
>
> As a result of the economic crisis, cash and liquidity management ranked among the top priorities of this company's CFO, so the CIO outfitted the next-generation executive information system with a 'flexible periphery' linked to financial markets and other data sources. Executives can now model the way market movements influence the cost of capital and cash flows.The remaining tiers house the supporting software and data structures (Layer 4a) and the IT infrastructure, with hardware and networks (Layer 4b). A service-oriented Layer 3, structured by the most important business domains, aligns business requirements with IT capabilities to ensure that the executive information system tags, synthesizes, and applies data efficiently.

The tiered structure proposed in the above example is a robust way to provide an information resource arrangement to monitor important KPIs, and to drill down into detail or join with other data and use analytic functions whenever deeper sense is required. Such arrangement conforms well to a tiered structure.

The two-tier (or more) structure of user support environment might assist in balancing the demand-driven and supply-driven informing approaches in a way that common needs, served by repeating routine procedures, are mostly satisfied in the first tier of this environment, with a probable exception of common complex needs—e.g., discovery of hidden rules in big data or monitoring the environment for emerging innovations. The emergence of special needs, requiring additional attention, might require more support resources than first tier can provide, and thus facilitate the use of the additional resources found on second tier or further tiers, if they do exist. Special needs do not always need second tier "heavy artillery"—in some cases the more common tools of the first tier might suffice. Same applies to the common needs—in some cases more sophisticated techniques of the second tier may be necessary—e.g., for advanced evaluation of a general view.

4.7 The Encompassing Potential of BI

The coverage of business dimensions by BI dimensions is mostly important in creating an encompassing picture of the business and its environment, while at the same time focusing on the most important business processes, directions and risks. While discussing data warehouses as one of the possible forms of accumulating intelligence data, March and Hevner (2007) argue that data warehouse should not just serve as a repository of internal operational data, but also cover partnerships, business rules, competitors and markets, goals, opportunities and problems, successes and failures. Put differently, such data warehouse should cover as many **key business dimensions** as possible with a potential goal of developing a balanced holistic view. In other words, this understanding of data warehouse is shifting closer to a notion of global business data collection, which is not entirely structured—a significant part of it is unstructured or semi-structured. March and Hevner also note that methodologies and representational formalisms for encompassing analysis in such cases are seriously lacking.

The importance of supporting multidimensionality is well rounded up in (Choo, 2016): "... The Intensified Smallpox Eradication Program started by setting goals and strategies that were based on the best evidence available at the time. Over time, those who carried out the program learned that knowledge is provisional and always being improved upon; that beliefs should be linked to practical experience; that learning is a distributed, community-based activity; and that the acquisition of knowledge and information is embedded in a context of norms, rules and traditions that defined the institutions and communities participating in the program. The locus of learning thus alternated between the global and the local, between individual discovery and collective inquiry, and between active problem-solving and reflexive evaluation of beliefs."

In practical situations, the required information and coverage is seldom complete because of resource limitations, and a portfolio principle is likely to be used to use the available resources—people, time, money—in a best possible way.

References

Abbasi, A., Albrecht, C., Vance, A., & Hansen, J. (2012). MetaFraud: A meta-learning framework for detecting financial fraud. *MIS Quarterly, 36*(4), 1293–1327.

Anderson-Lehman et al. (2004). MISQ—/Implem.

Antova, L., Baldwin, R., Gu, Z., & Wass, F. M. (2015). An integrated architecture for real-time and historical analytics in financial services. In M. Castellanos, P. K. Chrysanthis, & K. Pelechrinis (Eds.), *Real-time business intelligence and analytics: International workshops 2015–2017. Revised selected papers.* Cham, Switzerland: Springer.

Berkman, E. (2001). When bad things happen to good ideas: Knowledge Management is a Solid Concept the Fell in with the Wrong Company: Software Companies, to be Precise. *Darwin Magazine, 2001,* 51–56.

Boyer, J., Frank, B., Green, B., Harris, T., & Van De Vanter, K. (2010). *Business intelligence strategy: A practical guide for achieving BI excellence*. Ketchum, ID: MC Press Online.

Bucher, T., Gericke, A., & Sigg, S. (2009). Process-centric business intelligence. *Business Process Management Journal, 15*(3), 408–429.

Business Health Check. (n.d.). *HTC consulting*. Retrieved January 15, 2019, from http://www.htc-consult.com/Healthcheck.htm.

Carlo, J. L., Lyytinen, K., & Boland, R. J. (2012). Dialectics of collective minding: Contradictory appropriations of information technology in a high-risk project. *MIS Quarterly, 36*(4), 1081–1108.

Choo, C. W. (2016). *The inquiring organization: How organizations acquire knowledge & seek information*. Oxford: Oxford University Press.

Cohen, W. M., & Levinthal, D. A. (1990). Absorptive capacity: A new perspective on learning and innovation. *Administrative Science Quarterly, 35*, 128–152.

Daly, M. (2016). Decision support: A matter of information supply and demand. *Journal of Decision Systems, 25*(Suppl), 216–227.

Davenport, T. (2015). The insight-driven organization: Management of insights is the key. *Deloitte Insights*, August 26, 2015. Retrieved August 23, 2018, from https://www2.deloitte.com/insights/us/en/topics/analytics/insight-driven-organization-insight-management.html.

Dresner, H. (2017). *Advanced and predictive analytics market study*. Dresner: Advisory Services LLC..

Dwivedi, J. (2016). *Creating a data-driven culture (comment)*. Retrieved January 3, 2020, from https://www.ngdata.com/creating-a-data-driven-culture/#Dwivedi.

Dykes, B. (2016). *Actionable insights: The missing link between data and business value*. Retrieved December 1, 2017, from https://www.forbes.com/sites/brentdykes/2016/04/26/actionable-insights-the-missing-link-between-data-and-business-value/#48e0ef2251e5.

Eckerson, W. (2013). *Organizing the BICC Part I: Move to the middle*. Retrieved January 15, 2018, from http://www.b-eye-network.com/blogs/eckerson/archives/2013/10/the_secret_to_o.php.

Elbashir, M. Z., Collier, P. A., & Davern, M. J. (2008). Measuring the effects of business intelligence systems: The relationship between business process and organizational performance. *International Journal of Accounting Systems, 9*, 135–153.

Elliott, T. (2004). *Choosing a business intelligence standard. Business objects white paper*. Retrieved June 3, 2020, from https://silo.tips/queue/white-paper-choosing-a-business-intelligence-standard?&queue_id=-1&v=1597662499&u=NzguNTYuMTY5LjY=.

Elliott, T. (2014). *5 Top tips for agile analytics organizations*. Retrieved June 12, 2020, from https://timoelliott.com/blog/2014/09/5-top-tips-for-agile-analytics-organizations.html.

Gilkey, Ch. (2011). *The four key dimensions of business*. Retrieved January 30, 2019, from https://www.productiveflourishing.com/the-four-key-dimensions-of-business/.

Hackathorn, R. (2004). The BI watch: Real-time to real-value. *DM Review*, January 2004.

Harris, Ph., Hendricks, M., Logan, E.A., & Juras, P. (2018). *A reality check for today's C-suite on Industry 4.0. KPMG International*. Retrieved January 30, 2019, from https://assets.kpmg/content/dam/kpmg/xx/pdf/2018/11/a-reality-check-for-todays-c-suite-on-industry-4-0.pdf.

Hasan, S., O'Riain, S., & Curry, E. (2012). Approximate semantic matching of heterogeneous events. *Proceedings of the 6th ACM International Conference on Distributed Event-Based Systems*, July 16–20, 2012, Berlin, Germany. pp. 252–263.

Hernandez, J., Berkey, B., & Bhattacharya, R. (2013). *Building an analytics-driven organization. Accenture*. Retrieved July 22, 2020, from https://www.accenture.com/us-en/~/media/accenture/conversion-assets/dotcom/documents/global/pdf/industries_2/accenture-building-analytics-driven-organization.pdf.

Huwe, T. (2011). Meaning-based computing. *Online, September–October 2011, 35*(5), 14–18.

Kamal, M. M. (2012). Shared services: Lessons from private sector for public sector domain. *Journal of Enterprise Information Management, 25*(5), 431–440. https://doi.org/10.1108/17410391211265124.

Kasibhatla, K. N. (2017). What's holding you back? How to build an insight-driven organization. Forbes, December 18, 2017. *Interactive*. Retrieved March 1, 2018, from https://www.forbes.com/sites/forbestechcouncil/2017/12/18/whats-holding-you-back-how-to-build-an-insight-driven-organization/#4ea32c9c310a.

Krensky, P. (2015). Self-service analytics, the cloud, and timely insight. *Aberdeen Group report*. Retrieved July 22, 2020, from https://docplayer.net/23731971-Self-service-analytics-the-cloud-and-timely-insight.html.

Liautaud, B., & Hammond, M. (2001). *E-Business intelligence: Turning information into knowledge into profit*. New York, NY: McGraw-Hill.

Luckham, D. C., & Frasca, B. (1998). Complex event processing in distributed systems. In *Computer systems laboratory technical report CSL-TR-98-754*. Stanford: Stanford University.

Lynch, T., & Gregor, S. (2003). Technology-push or user-pull? The slow death of the transfer-of-technology approach to intelligent support systems development. In S. Clarke, E. Coakes, H. M. Gordon, & A. Wenn (Eds.), *Socio-technical and human cognition elements of information systems*. Hershey, PA: Idea Group Publishing.

March, S. T., & Hevner, A. R. (2007). Integrated decision support systems: A data warehousing perspective. *Decision Support Systems, 43*, 1031–1043.

Mayer, S. (2010). *One company's blueprint for a next-generation executive information system*.

Meredith, R., Remington, S., O'Donnell, P., & Sharma, N. (2012). Organisational transformation through Business Intelligence: Theory, the vendor perspective and a research agenda. *Journal of Decision Systems, 21*(3), 187–201.

Miller, J. (2000). Millenium intelligence. In *Understanding and conducting competitive intelligence in the digital age*. Medford, NJ: CyberAge Books.

Ramakrishnan, T., Jones, M. C., & Sidorova, A. (2012). Factors influencing business intelligence (BI) data collection strategies: An empirical investigation. *Decision Support Systems, 52*, 486–496.

Ross, S. (2001). Get Smar—Looking for diamonds in data? With business intelligence software, you may find buried valuable nuggets and stop being taken by the fool's gold. *PC Magazine*, August 2001.

Skyrius, R. (2008). The current state of decision support in Lithuanian business. *Information Research, 13*(2), 345. Retrieved March 1, 2018, from http://InformationR.net/ir/31-2/paper345.html.

Skyrius, R. (2014). The split of information needs support between the users and information technology. In G. Phillips-Wren & S. Carlsson (Eds.), *DSS 2.0—Supporting decision making with new technologies: 17th conference for IFIP WG8.3 DSS, 2–5 June 2014, Paris, France: Supplemental proceedings* (pp. 227–238). Amsterdam: IOS Press.

Skyrius, R. (2015). The key dimensions of business intelligence. In K. Nelson (Ed.), *Business intelligence, strategies and ethics* (pp. 27–72). Hauppauge: Nova Science Publishers, ISBN: 78-1-63482-064-6.

Skyrius, R., & Bujauskas, V. (2011) Business intelligence and competitive intelligence: Separate activities or parts of integrated process? *The global business, economics and finance research conference*. London, UK, July 2011.

Skyrius, R., & Nemitko, S. (2018). The support of human factors for encompassing business intelligence. *Proceedings of Informing Science & IT Education Conference (InSITE) 2018*, 21–34. Retrieved January 12, 2019, from http://proceedings.informingscience.org/InSITE2018/InSITE2018p021-034Skyrius4527.pdf.

Skyrius, R., & Winer, C. R. (2000). IT and management decision support in two different economies: A comparative study. In *Challenges of information technology management in the 21st century: 2000 information resources management association international conference*. Alaska, USA: Anchorage.

Snowden, D., & Boone, M. (2007). A leaders' framework for decision making. *Harvard Business Review, 85*(11), 69–76.

Standen, J. (2010). *Let's admit it—Centralized business intelligence alone just doesn't work.* Retrieved January 30, 2019, from http://www.datamartist.com/centralized-business-intelli gence-alone-does-not-work.

Stangarone, J. (2015). *7 Practical ways to improve user BI adoption.* Retrieved January 31, 2020, from https://www.mrc-productivity.com/blog/2015/03/7-practical-ways-to-improve-bi-user-adoption/.

Straub, D., & DelGiudice, M. (2012). Use. *MIS Quarterly, 36*(4), iii–viii.

The 8 Key Dimensions We Evaluate When Assessing a Company's Growth and Innovation Capabilities. (2015). *Growth Decisions.*Retrieved January 30, 2019, from https://www. growthdecisions.com/archive/2015/3/9/the-8-key-dimensions-we-evaluate-when-assessing-a-companys-growth-and-innovation-capabilities.

Tyson, K. (2006). *The complete guide to competitive intelligence.* Chicago, IL: Leading Edge Publications.

Vuori, V. (2007). Methods of defining business information needs. In *Frontiers of e-Business research conference.* Tampere, Finland: Tampere University of Technology.

Walker, R. (2009). *The evolution and future of business intelligence.* Retrieved January 9, 2011, from http://www.information-management.com/infodirect/2009_140/business_intelligence_bi-10016145-1.html.

White, C. (2007). *Who needs real-time business intelligence? Teradata Magazine, September 2007.*

Yeoh, W., & Coronios, A. (2010). Critical success factors for business intelligence systems. *Journal of Computer Information Systems, 50*(3), 23–32.

Chapter 5
Information Integration

5.1 The Need for Data and Information Integration

Unlike traditional management information systems, who operate in fairly defined functional areas, business intelligence often is expected to provide a *composite view of a field of interest,* where relevant business dimensions are joined together to produce adequate coverage and context. However, most organizations, who use advanced BI tools and employ smart people, never produce a *complete picture* that they require (Huret, 2018). The achieved result is more of a partial interpretation of an incomplete picture, and the main reason for this is the scattered data sources and repositories that were built at different times for different reasons. The problem is compounded internally by the existence of data silos around functional areas, although contemporary ERP-type systems are intended to ensure integration of internal data. Regarding external environment, data and information are collected from various channels in assorted formats, produced using different conventions and possessing uneven reliability, and often hard to combine.

Attention to information integration soared in the first decade of XXI century using terms like EII (Enterprise Information Integration), EAI (Enterprise Application Integration), enterprise portals etc. Traditionally such approaches have been predominantly technology-centric, as some sources stated (e.g., Roth, Wolfson, Kleewein, & Nelin, 2002), rarely mentioning the final goal of the users, and leaving unclear the expected value to justify huge investments and effort. The notion "the more information you have, the better" has been replaced by "the more information you integrate, the better", but somehow the user needs and business cases are still not the principal starting point for information integration issues (Loshin, 2013).

Integration value: The understanding of the environment is usually insufficient; information integration is intended to provide:

- **Bigger or wider picture:** Uniform interface to many data sources (Alevy, in Hearst, Levy, Knoblock, Minton, & Cohen, 1998); combination of data from

© Springer Nature Switzerland AG 2021
R. Skyrius, *Business Intelligence*, Progress in IS,
https://doi.org/10.1007/978-3-030-67032-0_5

independent sources; an ability to see more complete picture by covering gaps; faster and more confident actionable intelligence because of quality integration;

– **Deeper insight:** deeper understanding of the topic when integration encompasses relevant information; discovery of hidden dependencies and rules—not only in Big data, but in other information resources as well; deeper insights and explanations (Cody, Kreulen, Krishna, & Spangler, 2002); enterprise information fusion combining internal and external sources (Shroff, Aggarwal, & Dey, 2011).

The need for information integration, as stated in (Skyrius, Šimkonis, & Sirtautas, 2014), is supported by two groups of important factors:

- Factors of expected value—**deficit** of relevant and meaningful information; the hidden value "buried" in the organization's information assets or its environment;
- Obstacles for simple extraction of this value—**overload** (glut) with data and irrelevant information at the same time; attention-hungry integration procedures are complicated by the variety of sources, formats, agreed meanings and signs etc.; emergence of Big Data; growth of importance of unstructured information; growing needs to automate information integration;

The results from BI activities—developed insights and new useful information—have to be *actionable* to utilize their potential, and many sources agree that to be truly actionable, this view is expected to be as encompassing and reliable as possible. Kohavi, Rothleder, and Simoudis (2002) have stated that integration of both data and information helps boosting the actionable qualities of the results of BI activities. For instance, integration of operations and analytics systems improves the collect-analyze-act-measure cycle. Following this logic, integration of data and information from different types of operations may assist in discovery of important signals indicating either growing problems or opening opportunities, and point more clearly to the need for action, its type and urgency.

Many academic and practical sources assign different meanings of the term "information integration"; several more common examples are:

- Introduction of common data standards;
- Data loading into a data warehouse;
- A federated join across independent data sources;
- Place together bits and pieces about a behavior of a competitor;
- Monitor competing products on the market;
- Integrate supply chain data to harmonize processes;
- Do a thorough due diligence on a merger/acquisition prospect;
- Behavior of a product;
- Behavior of a customer segment; etc.

While declaring they are researching information integration issues, the sources actually take one of the two paths: *data integration* or *information integration*. Data integration revolves mostly around technical problems, and information integration follows different principles—relations are created around a certain axis topic, and related by tokens of sense, not data serving as keys between tables or other data

Table 5.1 The scale of integration complexity levels

	Features	Example
Simple integration	Integrates structured data by rules developed in database domain. Data carry elementary (atomic) sense that is well-known beforehand. Key issues: data quality, coding and other conventions	Joining data on the same objects from different sources—e.g., "all data on this customer"; "lost shipment" etc. A record of sale to a customer joined to customer status, or age, or address, or other demographics of this customer EII and ETL procedures
Medium complexity integration	Integrates structured and unstructured information by fragments of sense that have to be extracted (IT function) and related to other information (mixed IT-human function). Annotations and tagging are among the most common tools	Text mining with key terms or phrases. Extraction of structural elements from unstructured or semi-structured sources: names, dates, codes etc. Inquiry for research papers with a specific term, and most used other terms in text close to the primary term Automatic classification of documents or research papers. Aggregation of available information on a key competitor
Complex integration	Integrates information on the basis of sense-making. Assisted by the above types of integration, but carries on from complex meanings delivered with information from various sources to develop required insights or answer complicated questions	Evaluation of competitor strategy Evaluation of a prospective and possibly disruptive innovation Prospects of entering a market. Detection of a major oncoming crisis

Source: author

collections. Such sense-handling issues are significantly more complicated to support by IT resources. This chapter intends to have a closer look at the existing approaches to both data integration and information integration.

Data integration is usually seen as a simpler task, however complicated technically it may be, and if a continuum of simple to complex integration can be drawn, information integration is to be placed on the complex end. A scale of integration complexity levels (Table 5.1) would start with simple integration or no integration at all, and end with complex information integration cases where multiple and often conflicting meanings and contexts have to be remedied to achieve satisfactory insights.

The above scale of integration complexity levels mirrors the scale of simple-complex information needs, presented in the Chap. 3 of this book. Another dimension of common-special information needs may also be related to different cases of information integration:

1. An example of **common integration:** monitoring of all competition in the market by combining sales, financial and other available data. The result of such integration is a composite evaluation of a behavior of a group of objects of interest.
2. An example of **special integration**: a thorough monitoring of a certain competitor to predict possible moves and prepare counteraction, if required.

All the above examples aim towards a certain level of sense-making, yet differ in their scope, context, attention focus point. The remaining part of this chapter will look at several more common issues of integration, starting with the simple level of data integration.

5.2   Data Integration

Data integration, while dealing with atomic units of sense, produces enrichment of context around issues of interest, moves perception of a situation towards a more encompassing view, and plays an important role in insight building. The importance of data integration shows across all activities and industries—for instance, Schneider and Jimenez (2012) provide solid arguments for data integration in life sciences, where value is created by providing extra scope, and thus more potential for new insights. Given that business intelligence uses many discovery techniques borrowed from scientific research, this is hardly surprising.

The examples of data integration are abundant: in retail, data integration across outlets and channels helps answering questions like "which locations or processes are under-performing? What is the inventory of a certain item across storage locations? Which items or groups experience the most often shortages?" In client relations, data integration helps estimating what are the most promising campaigns based on behavior rules for certain groups of customers, or revealing what behavior is signaling that a client is about to defect to a competitor. In healthcare, the axis is a patient; complete picture on similar cases allows discovering many important trends. Although the analytical principles in healthcare do not differ much from those used in marketing, the stakes are different—a fast classification of an emergency case can save lives (see emergency room data mining example in (Portela, Aguiar, Santos, Silva, & Rua, 2013)). In finance, fraud detection may unite data on many similar-looking cases, eventually distilling important features and building profiles; AI-based methods can be used to recognize fraud, and, if possible, proactively intervene.

The absence of data integration produces "business information gaps" that can be the roots of many business problems. Several examples are presented below (Williams, 2016):

- Marketing—incomplete view of shipments across channels; or return on campaigns; or management of brand portfolio; or management of inventory levels;
- Sales—access to key performance measures; lack of historical performance information; inability to leverage dashboards to manage by exception; lack of single

source of truth (SST); no consistent access to current sales and inventory levels; no ROI on promotional campaigns; no efficient way to measure and manage in-store execution of promotions; overall difficult measure of KPIs;

- Finance—difficulties in managing drivers of plant profitability; lack of standardized historical performance information; lack of analysis in budget variances; outdated spreadsheet-based COGS.

Verhoef, Kooge, and Walk (2016) present several examples of user comments that reflect the needs for data integration:

- We have tons of data, so why is it taking so much time to create the right insights when we need them?
- Why do we have to gather our crucial marketing insights from so many different departments within the organization?
- We are overloaded with reports and overviews, but they don't give us input on how to improve our business performance.

Much of data integration issues revolve around use of unified coding and semantic consistency. For these purposes, coding conventions are developed and followed on organizational, national or international levels. However, officially accepted coding conventions may mismatch the actual required points of disclosure—e.g., aggregation levels are too high, or too low, or have different logic than emerging business problems are asking for.

For overcoming compatibility and consistency obstacles, XML (Extensible markup Language) is an often-used data integration standard that has a tag-driven flexible structure. This structure allows substantial flexibility in relating data from sources that are incomplete, in different formats or with different data schemas.

An approach of data spaces had been proposed by Franklin, Halevy, and Maier (2005), who stated that "developers are more often faced with a set of loosely connected data sources and thus must individually and repeatedly address low-level data management challenges across heterogeneous collections." The proposed data spaces would support data co-existence instead of rigid integration, providing a flexible and holistic data view. The current spread of the notion of data lakes may be seen as a step in the same direction: a data lake is expected to accept all types of data, perform the minimal data transformation or no transformation at all; and provide the necessary schemas for analysis.

In an ideal case of data integration, according to Haas (n.d.), most of integration development including discovery of data, metadata, schemas, engines, should happen automatically, and the user should see no visible difference between integration methods in operation. The available experience in data integration might suggest that such seamless development and operation is more expected from "loose" approaches to data integration (data spaces, data lakes, Hadoop framework) than from "tight" approaches (data warehouse). Consequently, the expected agility and longevity of integration approaches would depend on flexibility of underlying methods.

An important factor in data integration is the existence of data silos that exist as data collections organized by separate functions or departments inside an organization. While data separation by subject area satisfies many local information needs, the isolation of such collections from the rest of the organization leads to lack of transparency and efficiency, and, from BI point of view, prevents the production of a "big picture". Data silos make collaboration suffer, slow down business advance, and create data synchronization issues because of possible data duplication.

Amaresan (2019) provides several factors that drive silo formation:

1. Organization growth and size. For large, and especially global, organizations, smooth information sharing may be not easy at all, and internal competition may prevent from granting access to the data of a separate "county". We can add that such silos may be run on local systems that are proven reliable, so changes would definitely meet resistance.
2. Technology differences. Smooth data flow across functions requires similar or compatible technology, which is not always the case.

Data silos are part of the larger centralization-decentralization tradeoff, concerning data location and access rights. One of the virtues of centralization is the conformance to general standards, leading to encompassing views and wide visibility. A technology of central data warehouse enables a highly centralized data integration solution, and data have to be highly standardized and organized to reap the possible benefits. Naturally, such approach imposes certain limitations in terms of flexibility and data inclusion, leading to the search of opposite solutions that are less demanding in terms of standards and organization.

Holman (2013) has proposed a cyclical structure for data integration, defining data integration as a process with constant feedback and improvement (Fig. 5.1):

The above cycle roughly conforms to the ETL (Extract, Transform, Load) process structure, which is discussed in more detail in Chap. 7—Technology chapter. The exception in the above cycle is the absence of the L (Load) phase that is specific for data warehouses or other collections intended to create one more instance of source data. Instead, the above cycle uses Delivery phase, which unites data from original sources in a some kind of federation model without creating a new data instance. The closed-loop format, used in several places in this book, although rarely associated with ETL-type processes, reflects the importance of feedback and lessons learned for next iterations of the same process.

5.2.1   Single Source of Truth

Regardless of data organization model, any organization as a single entity needs to maintain consistency in informing activities. The complexity of business information flows grows exponentially with constantly changing products, customers, suppliers, further complicated by significant changes like mergers, alliances or relocations. For a complete and reliable coverage, an important issue, especially for large organizations, is semantic heterogeneity (Halevy et al., 2005) that justifies the need for a specific reference tool called SST (Single Source of Truth). This need is often the driving factor to launch a BI project.

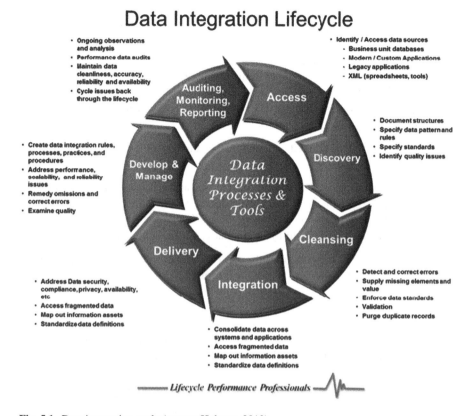

Fig. 5.1 Data integration cycle (source: Holman, 2013)

SST is often defined as a single instance or single definition of data, where all the other uses are through reference. Some sources (Grealou, 2016) define these two issues more precisely:

- Single source of truth is a data storage principle to have a particular piece of data in one place;
- Single version of truth is a view of data that is agreed by everyone concerned to be the trusted definition of a piece of data.

Of the two, the second issue is less technical and more dependent on a definition. A definition may regard to a generic, and often vague, term—lead, client, project, account. Loshin (2013) accentuates a necessity to start with clearly defined business cases or questions, followed by *forced semantic consistency* through collaboration and agreements about the definitions for potentially ambiguous business terms. Some more common directions to collaboratively work out are:

- Identify key business terms and phrases, such as clients, projects or leads;

- Document contextual uses for the terms and phrases (what exactly does a "client", "project" or "lead" mean?);
- Introduce unified coding, definitions of meaning, relations—e.g., a project is owned by one department or more?
- Infer contextual meanings for the terms and phrases;
- Facilitate collaborative interactions for documenting and reviewing candidate definitions for the business glossary terms within each specific context;
- Foster agreement about the definitions.

Verhoef et al. (2016) pointed out analytical challenges of integrated data that lead to the necessity of SST.:

- Alignment of definitions of atomic objects—customers, consumers;
- Alignment of segmentations by specific dimensions—time, place etc.;
- Alignment of insights to be generated;
- Alignment of aggregations to maintain clarity for modeling purposes.

The main expected benefits from the use of SST lie in the areas of data quality, ease of understanding and accessibility of information.

5.2.2   Enterprise Information Integration and Enterprise Application Integration

EII (Enterprise Information Integration) and EAI (Enterprise Application Integration) were born in the first part of the first decade of twenty-first century out of concerns about integrating company data without actually loading it into a data warehouse. Both EII and EAI are somewhat outdated and widely criticized concepts; yet both of them have demonstrated certain integration rationale. Actually, the core idea of EII has been not information integration, but data integration. The approach is based on an idea to join data from heterogeneous sources by creating data views. Speed and simplicity are the main concerns, often at the expense of correctness and verifiability. EAI is expected to connect applications to communicate and support processes or workflows by using semantic tools like ontology languages, and is clearly oriented towards information integration, which is covered in one of the following paragraphs of this chapter.

Various criticisms have been addressed towards EII and EAI. Hayler (2004) has heavily criticized EII, describing it as cumbersome, inflexible and overly optimistic. The artificial separation of data integration into EII and EAI has been criticized as confusing and painful; simple read-only transactions are performed without smoothly, but data update issues are much more complicated. As a partial remedy, reinvention of both approaches into a single integration solution had been proposed (Carey, in Halevy et al., 2005). The very terms of EII and EAI are past their hype in the technology news sources, but the need for data and information integration will obviously remain.

Andriole (2006) describes several options for EAI, ranging from custom-developed "glue" to connect applications of assorted provenance, through exploiting some general interfaces built into popular applications, to generic "glue" that was designed to connect lots of applications and databases. However, it has to be noted that even the most standard and generic connections need tweaking to make them work in a real environment. Probably it is important to realize that custom integration connections that have solved a data integration problem in the past may actually add to the fragility of the whole system by making it strained and rather vulnerable to any changes like technology upgrade.

The issue of future flexibility, according to Bilhorn and Hulsebus (2014), is addressed not so much through direct flexibility of APIs and field mapping, as through mapping processes and people who are the integration clients. A thorough inquiry into client expectations would reveal the key points of agility that would reduce the pain of future changes.

5.2.3   Data Quality

In business intelligence, as in any informing activity, data quality is a huge issue that stretches well outside the scope of this book. A bitter fact is that, according to Park, Huh, Oh, and Han (2012), academics and researchers have paid little attention to the issues surrounding data quality validation. BI initiatives are only capable of optimal performance when the data utilized are of high quality (Redman, 2004; Wang & Strong, 1996). Although well-designed and highly functional, BI systems that include inaccurate and unreliable data will have only limited value, and can even adversely impact business performance.

5.3   The Transition from Data to Information Integration

Although acting towards the same goal—supporting the development of business insights, data integration is executed from formal systems and deals mostly with strong signals, and information integration seeks making sense from variety, chaos, ambiguity and weak signals. The two are often combined—let's say, a problem requires monitoring social media and public sources; the results are merged with financial documents and analyst reports from other channels; complex natural language processing identifies specific references that are organized by axis—time, topics, concepts; finally visualization is used to point out the most important trends. According to Ananthanarayan, Balakrishnan, Reinwald, and Yee (2013), information integration from heterogeneous sources is one of the major challenges facing businesses. Specifically, integration of structured and unstructured data is of growing importance because of search for more complete insights on important business aspects, as well as unstructured data becoming more and more pervasive (Bernstein & Haas, 2008; Doan et al., 2009).

Roth et al. (2002) define information integration as "a technology approach that combines core elements from data management systems, content management systems, data warehouses, and other enterprise applications into a *common platform*". This definition has been proposed way before the mention of data spaces and data lakes. However, it expresses a limited technology-centrist view; this is not exactly information integration—combining onto a common platform is just the beginning, and handles the technical preparation phase. This has to be followed by procedures of different nature, dealing with extraction of information and integration of sense. In technology-centric sources, the process of information integration would roughly consist of two phases:

1. Extract structure from unstructured information (names, technical terms); find and save dimensional elements—codes, dates, addresses etc.
2. Place discovered structural elements into a separate entity, define/discover relationships, extract and link meanings. Validity check between discovered elements—e.g., if the same object shows up in contexts that are mutually exclusive—while potentially useful, is complicated.

Data to information (structured—unstructured). Integration of unstructured data is important and largely under-researched activity, requiring different approaches and bordering with information integration. Unstructured data comes in the form of text, pictures, video, and often requires intelligence processing methods—computational linguistics, image recognition and similar. Basically, it is a task of recognition and attribution—atomic features of the objects of interest are recognized that serve as identifiers. There are more advanced tasks of information integration—for example, production of a summary of several documents, where one has to deal with higher levels of abstraction.

There's an often-mentioned statement regarding the share of unstructured information in the information resources in an organization, stating that this share is around or no less than 80% of the total. It has to be said here that this figure has never been certified by any research, and it is heavily criticized in many sources, e.g., Grimes (2008), White (2018). The actual research provides quite different figures—a TDWI report quotes 47% of structured data, 22% of semi-structured data, and 31% of unstructured data (Russom, 2007).

It is interesting to note that in certain cases of information integration separate pieces of information are connected not directly through sense tokens (this text is about X, and this note is about Y, but also about X, so they may be related), but through assigned formal data items—codes, labels and the like. An example is text analysis software that recognizes identifiers in unstructured text—names, phone numbers, zip codes, email addresses etc.—and assigns them to some created header that is attached to the text; unstructured data becomes partly structured. Such case may be labelled "hard" integration that is supposed to be performed by matching sense tokens or some structured data, often by using specific software. An example of soft integration could be the so-called information basket analysis, done along the same principles as basket analysis in retail data. Just like basket analysis in retail estimates frequent combinations of goods purchased together, information basket analysis looks at cases where seemingly independent pieces of information are required to be presented together and joined by the user if a relation is noticed. Some data from earlier research on combinations of data types is presented close to the end of this chapter.

5.4 Information Integration and Sense Making

While many issues in this subchapter are of considerable width and have merited thorough research, here they are presented as influencing information integration issues important in BI processes.

5.4.1 Insight Development and Information Integration

While many sources admit that information integration is performed in search for rich, aggregated meanings, the very notion of information integration is somewhat bloated. Information integration is often seen as a series of procedures performed to develop awareness—information search and collection, evaluating, joining in a schema or model, gaining a new level of understanding in the case of success and possibly leading to insight development. Pirolli and Card (2005) define information integration as a multi-step process with recurring loops in every step, having opportunistic nature. The higher complexity of a problem, the more the nature of linking sense will shift from data integration to information integration, handling "soft" objects carrying relevant information—ideas, concepts, judgments, comments etc.

Salmen, Malyuta, Hansen, Cronen, and Smith (2011), while addressing defense intelligence issues, noted that traditional data integration approaches fall short when facing heterogeneous sources with diverse structure, scale and modality. Several important obstacles emerge: need for heavy pre-processing; difficult integration of data models for different data sources; possible loss or distortion of data, associated semantics or provenance due to diverse modality; different analysts may have different points of view on data to be integrated, or use different analytical approaches and tools. Information integration process deals with incomplete and uncertain information, is extremely knowledge-intensive and expensive in terms of human costs (Levy, 1998). Barrett et al. (2011), discussing information integration in networks to assist policy decisions, stressed complex interaction in networks of elements from diverse yet connected areas. Writing about policy problems, authors state that "integration of information is a particularly challenging problem, especially in the face of massive data sources that have been collected by different individuals and institutions in parallel, and rarely specifically for the issue of interest". ... The paper also gives a rough framework for a solution supporting policy decisions: such solution has to provide:

• Support for multiple views and optimization criteria for multiple stakeholders (adaptability);
• Ability to include multiple sources of data (extensibility);
• Capability to model very large interacting systems (scalability); and
• Support for evaluation of numerous policy alternatives.

An example of information integration around a clear axis from the area of management decision support is presented in (Saunders & Jones, 1990), and describes a development of a decision for supercomputer acquisition. A set of assorted informing procedures is executed until it is agreed that the issue is clear enough, or no further useful arguments are expected, or time and money limitations are pressing. These procedures add a series of information snippets to the available body of information on the subject—feasibility of purchasing a Cray supercomputer for complicated oil field drilling simulation tasks. The information is from various sources and in various formats, from cost-benefit analysis models to phone calls and grapevine. Some of it confirms earlier assumptions, while other information reduces their value. However, uncertainty is reduced with almost every step.

Information integration advances towards an insight on a particular issue. Many sources see this advance as a process of sequential steps with possible recursion and iterations (Pirolli & Card, 2005). There are views that oppose the structured sequential nature of information integration process: Klein (2015) discussed the predictability of the insight development process, and argued that in practice decision making and insight development are far from being rational:

– "We no longer claim that the only way to make a good decision is to generate several options and compare them to pick the best one (experienced decision-makers can draw on patterns to handle time pressure and never even compare options);
– We no longer believe that expertise is based on learning rules and procedures (it primarily depends on tacit knowledge);
– We no longer believe that projects must start with a clear description of the goal (many projects involve wicked problems and ill-defined goals);
– We no longer believe that people make sense of events by building up from data to information to knowledge to understanding (experienced personnel use their mental models to define what counts as data in the first place);
– We no longer believe that insights arise by overcoming mental sets (they also arise by detecting contradictions and anomalies and by noticing connections);
– We no longer believe that we can reduce uncertainty by gathering more information (performance seems to go down when too much information is gathered—Uncertainty can stem from inadequate framing of data, not just from the absence of data;

– We no longer believe that we can improve performance by teaching critical thinking precepts such as listing assumptions (too often the flawed assumptions are ones we are not even aware of and would never list."

From both approaches we can draw that there are arguments against sequential information integration, but there are cases of a crucial piece of information that does the job of integrating information and triggering an insight; "the pieces finally fit", even for experts and gurus.

A good example of information integration is presented in (Miller, 2000) on pp. 166–167:

> When you combine the best practices and technologies from knowledge management and intelligence functions, interesting things start to happen. Different levels of expertise are accessed, new pieces of information present themselves, and observations start to aggregate and cluster, revealing new insights. Consider the following hypothetical example:
>
> One member of a knowledge network who works in the intellectual property division in New York City identifies a new patent in the area of digital imaging and holography. A second person in engineering in Silicon Valley locates a white paper published by Queen's College in London describing new laser technology. A third, working in Singapore, sees a notice in the local newspaper about the formation of a new corporation detailing its line of business and identifying its board members and investors.
>
> These three pieces of information are valuable, but think of the value if you aggregated them and discovered that one of the individual patent-holders is a co-author of the Queen's college white paper as well as a board member and investor in the newly formed Singapore corporation. If you happen to be in the business intelligence group at Eastman Kodak, for example, and your twenty-first century business plan is firmly routed in electronic imaging, you might be deciding whether to partner with or buy this company. Each item is interesting, but when you layer the items and look for common points of interest, the whole can become greater than the sum of its parts. By aggregating individual pieces of information and identifying common threads, you can create exceptional value.

Desouza and Hensgen (2005) present an example of failed information integration:

> Assume that three sources within an organization have data, collected separately, that, when combined, produce information of importance to the organization. Two weeks before a planned conference, one of the sources is burdened with an unrelated task, which draws it away from its initial work before the data can be used. The second source becomes the victim of company "rightsizing" and is let go. At this stage, the information opportunity to generate something that might have resulted from combining the data from the three sources has suffered the effects of maximum entropy.

The search for insights should be prepared for possible disappointments—the required information integration rarely happens under ideal conditions. We can present some examples from research on decision support information needs (performed in 2004), where the expected insights for decision making have been misleading or unreliable because of incomplete information and "white spots" in problem analysis. The below responses were the most common among 298 respondents:

1. If wrong decisions have been made because of lack of information, what kind of information?

   (a) Competitor action
   (b) Legal information
   (c) Informal confidential information
   (d) Incorrect aggregation, forecasting or other analysis
   (e) Specific; need not known in advance (emerged during decision making)
   (f) Client needs

2. Was this information possible to obtain?

   (a) Yes (in general)
   (b) No; confidential
   (c) Yes, with extra effort and using more sources (they may be distant and expensive)
   (d) Yes; human factor—incompetence and inadequate effort; underrated importance; accident;
   (e) Yes, but not on time
   (f) Only in live personal meeting

Looking at the above examples, we can first of all state that some of the vital information has been impossible to receive because of its confidential nature. The dominating responses regarding other cases mostly point towards human factors—incompetence, confusion, biased evaluation; and among all received answers, information technology or system guilt in lack of information is next to none. The above responses do not imply the need for some specific remedies to cover for the missing information. More likely, the responses point out that in most situations, especially more complicated ones related to decision support, incomplete information is a norm, and at the same time a practical problem that has been under-researched.

Having said earlier that information integration is always or most often centered on a certain **axis topic**, a couple of alternatives may be considered:

1. When there's a business issue, a problem or an opportunity, there's a central topic that serves as an integration axis: an object—a competitor or group of competitors; a customer or a group of customers; a supplier or a group of suppliers; a value of intended undertaking; a threatening regulation change; political turmoil etc. Looking at the BI dimension of data-driven versus issue-driven analysis, a presence of a clear (although not necessary true or final) topic places such analysis at the issue-driven part of the said dimension.
2. There may be cases with no clear problem at the start: detection of the most problematic or risky points from data analysis; discovery of odd events or anomalies that seem important; previously unknown links or relations. The monitoring function of the BI system is meant exactly for such goals. However, in the case of discovery a topic inevitably emerges, incurring special information needs, as opposed to common needs fulfilled by general monitoring.

One of the important issues in information integration, originating in knowledge management, has been the creation of common language (Dervin, 1998). Just as data integration needs a set of standards supported by, for example, SST repository, so does information integration by using standard vocabulary on some most important terms, especially those that are the likely candidates for integration axis—customers, projects, processes and so on.

5.4.2   Semantics, Sense Making and Context

In the area of information integration, one of the most important terms is "semantic", reflecting assigned meanings and meant to resolve semantic differences. Halevy, Norvig, and Pereira (2009) have pointed to a certain confusion around the term "semantic": "... Unfortunately, the fact that the word 'semantic' appears in both 'Semantic Web' and 'semantic interpretation' means that the two problems have often been conflated, causing needles and endless consternation and confusion. The "semantics" in Semantic Web services is embodied in the code that implements those services in accordance with the specifications expressed by the relevant ontologies and attached documentation. The "semantics" in semantic interpretation of natural languages is instead embodied in human cognitive and cultural processes whereby linguistic expression elicits expected responses and expected changes in cognitive state. Because of a huge shared cognitive and cultural context, linguistic expression can be highly ambiguous and still often be understood correctly."

As in data integration we need standard matching elements for any joins, in attempts to integrate information similar connecting elements are required—units of sense that are defined along some agreed conventions, and can be used to link snippets of information. Several types of such units have been proposed in published research, and are presented below:

1. **Ideas, notes, and comments** (Lau, Dimitrova, Yang-Turner, & Tzagarakis, 2014). While describing collaborative workspace, the authors have identified standard elements, or semantic types: ideas are items that deserve further exploitation (and trigger discussion); notes are additions of a more passive kind; and comments express less strong statements, are intended to be useful or explanatory or useful in some other way.
2. **Sign, concept, predicate, term, and statement** (Salmen et al., 2011). The paper proposes derived standard data model elements, or primitives—to introduce unified structure into source data, and to perform semantic enhancement. The product—enhanced ontologies, resolving of semantic conflicts, and support for information sharing.
3. Klein (2015) has pointed out what he had called five strategies for gaining insights:

   - **Connections**—a link that has not been known before;
   - **Curiosities**—odd events that open a discovery;

- **Coincidences**—a recurrent theme that draws attention and allows a discovery of an underlying cause;
- **Contradictions**—events that don't make sense; nevertheless, they are valid and force us to revise our existing beliefs;
- **Creative desperation**—getting stuck and then discarding some of older assumptions that are getting in the way.

Klein's list does not contain units in a sense of the first two cases, but they can be seen as derivatives or products of informing procedures, containing more complex sense aggregates than in the first two cases, together with *ad hoc* integrating procedures that produce unexpected results.

Although the above examples do not form a consecutive dimension of growing complexity of semantic units, they imply that automated definition and integration of complex information units is a daunting task, and probably should be best handled by joint human-machine effort.

In the recent history of knowledge management, the attempts to introduce units of measure have encountered criticism. For instance, Fahey and Prusak (1998) have pointed out that all such measures are surrogate, and orientate towards stock image of whatever is measured—knowledge or information—while giving minimal attention to the flow image that centers around movement or sharing of the above. Of course, information "on the move" (being read, transferred, received, interpreted) is much harder to measure that passive stock in some units, but, to author's opinion, this movement is where the value of information is redeemed.

Sense Making

For insight development, semantic information integration is essential to another key process—sense making. IT-based facilitation of sense making is a key point in this subchapter. As all issues regarding information integration are rather voluminous and extend well beyond the scope of this book, this subchapter is a brief overview of sensemaking in the context of information integration for business intelligence.

There are considerable advances of IT in the areas related to sense making—NLP, speech and image recognition based on machine learning techniques that keep rapidly developing their potential. However, in the dynamic business environment problems emerge and decisions are to be made under conditions of complexity and confusion, and development of reliable assessment rules seems to be a not-so-close prospect. At the time being the most rational approach seems to be what some sources call hybrid intelligence—the mix of human and IT-based intelligence competencies.

There are numerous definitions of sensemaking; we will quote here one made by Hutton, Klein, and Wiggins (2008) that provides a good average of most definitions: sensemaking is defined as the deliberate effort to understand events and is typically triggered by unexpected changes or surprises that make a decision maker doubt their prior understanding. Sense making is the active process of building, refining, questioning and recovering situation **awareness** by "joining the dots" or generating inferences, but also identifying what counts as a dot, and how to go about seeking new dots.

It can be noted that the ample literature on sensemaking does not exactly agree on the definition of its goal: while some sources state that sensemaking is intended to achieve full understanding of the issues at hand—"connect the dots", other sources discuss maximum possible understanding under given conditions. In any case, there seems to be a common agreement that sensemaking is never complete even in the cases of full understanding.

Karl Weick in his seminal work "Sensemaking in Organizations" (Weick, 1995), referred to seven properties of the sensemaking process. Although since the publishing of this book more sources on the same problem have appeared, these seven properties remain important in providing guidelines for better understanding of the sensemaking process. According to Weick, sensemaking is:

1. Grounded in identity construction because in responding to equivocal events, individuals and groups must determine who they are now in relation to a suddenly strange environment and who they will become as they start trying to change the environment.
2. Retrospective in nature because disruptions prompt individuals to turn their attention to information from the past in order to interpret how the current disruption came about.
3. Based on enacting sensible environments because a key output of sensemaking is an enacted environment that is more orderly than the equivocal environment that triggered sensemaking in the first place.
4. Social in that interpretations are negotiated and enacted through social interactions.
5. An ongoing process because sense is never made in perpetuity, but is always subject to disruption and therefore in need of re-accomplishment.
6. Focused on cues extracted from the environment because informational cues containing equivocality provide the raw material for interpretation.
7. Driven by plausibility rather than accuracy because sensemaking helps people reach only enough clarity to coordinate action, not to maximize expected outcomes with certainty.

Mudrik, Faivre, and Koch (2014) relate information integration and consciousness. Conscious experience is seen as **holistic**, undivided and combining multiple sources of information across space and time. For more complex cases, consciousness is required for integration over extended distances (SIW—spatial integration window) and time frames (TIW—temporal integration window), over higher semantic levels (SPIW—semantic processing integration window), for multiple modalities (MIW—multisensory integration window), and formation of novel associations with incongruent elements. The latter might be useful in dealing with strange, incongruent cues that might be a signal of something important.

A dichotomy of approaches to sense making can be noted here. Linear sensemaking approach scrutinizes available information on a problem, points out the dots to be connected, and finally connects them. Non-linear, heuristic, or ad-hoc approach is probably not so much about connecting the dots as recognizing the dots.

Of the sources advocating the linear sensemaking process, several are presented below. Pirolli and Card (2005) have examined the process of intelligence analysis, noting that it is largely a sensemaking task, and presented a model of analytical process, based on cognitive task analysis. The process covers several phases, organized into two major loops: a foraging loop, and a sensemaking loop. The steps of the process are joined by smaller cyclical loops, and there are two types of processes: bottom-up (data to theory) and top-down (theory to data). Bottom-up processes are of constructive nature, covering information search, extraction of useful snippets, schematization, case building and storytelling. Top-down processes mostly re-examine, re-evaluate, or suggest new directions. As the analytical process unfolds, simple information integration in a foraging loop gives way to advanced information integration in a sensemaking loop.

Weick, Sutcliffe, and Obstfeld (2005) characterized sense making as a process that is ongoing, subtle, swift, social, and easily taken for granted, and outlined the steps of sense making that are quite aligned with the steps of intelligence process:

- Sensemaking organizes flux from chaos;
- Starts with noticing and bracketing;
- Involves labeling and categorizing;
- Is retrospective;
- Sensemaking is about presumption and its testing;
- Is social and systemic;
- Is about action;
- Is about organizing through communication.

A related but different set of steps is presented in (Jennings & Greenwood, 2003), where rational activities are defined as an organizing process with three interlocking phases:

- Enactment—action that organizational members take as a result of interpretation of environment information that is puzzling or problematic;
- Selection—an interpretive process to reduce the number of possible meanings for information to become actionable;
- Retention—interpretations become organization's asset and reference points for future selection and enactment.

It is evident that some certain steps have to be present in the process of sensemaking—probably it should start with a **need** to make sense in situations whose understanding is unsatisfactory. The recognition of this need will likely be followed by the **assessment** of what is clearly known and what information relates to which aspect of the situation at hand. To cover for the missing information, some **assumptions** will have to be made. If the situation regards important issues, most likely it will have to be followed by **action**, at the same time experiencing

limitations of time and other resources. All the named steps will create **experience** which might be useful in the future. Once more, a cyclical process structure emerges, like for other processes in this book that are cognitive at their core.

Reflecting a different position, Lee and Abrams (2008) labelled sensemaking a complex **non-linear** process. For activities under turbulent conditions, there is an ongoing debate over whether organizations perform better engaging in thorough but lengthy information processing, or when they rely more on intuition and heuristics. Weick (1995) had noted that as turbulence goes up, so does the use of intuition and heuristics. An increase in complexity increases uncertainty because a greater number of elements create a greater number and variety of interactions. Madsbjerg (2017) defines sensemaking as "wisdom grounded in the humanities", differentiating it from structured algorithmic approaches. Klein (2013) presented a theory of sensemaking as a set of processes that is initiated when an individual or organization recognizes the inadequacy of their current understanding of events. Both definitions tend towards non-linear notion of sensemaking.

Proponents of both approaches—process and non-linear—have their arguments and, to author's opinion, the two approaches do not contradict each other. As stated by Weick et al. (2005), ". . . The order in organizational life comes just as much from the subtle, the small, the relational, the oral, the particular, and the momentary as it does from the conspicuous, the large, the substantive, the written, the general, and the sustained". Both approaches have a point where the emergence of insight is expected—the connection of dots, or the "aha!" moment. The process approach provides clearer structure that may give ground for expectations to attempt automation of sensemaking, while the fluid nature of non-linear approach suggests dominance of heuristic techniques like communication or brainstorming. Sense handling by IT is still in its infancy, and a purely algorithmic approach is hardly possible at the moment. Such issues as measurements and metrics of sense are significantly under-researched when compared to data engineering issues that are important in data integration. Based on the above, the potential of IT use for information integration seems to be on the side of hybrid approach, combining IT support with human heuristics.

Rounding up, as sensemaking does not limit itself to strictly sequential or loosely chaotic process, it is a blend of both modes. Some resources for this process are predefined, like background or context information, and more specific resources are harder to define or emerge in the process of sensemaking. IT support is expected to provide smooth, orderly informing for both of the above modes: simple information integration for the sequential part (collection, organizing, sorting, filtering, visualization) and more complex aids for the heuristic, loosely organized part (recall, relating, detecting, re-examining, re-evaluating) that would support fresh angles in assessing the situation at hand.

The Role of Context

Information integration creates and expands context for any business situation. Regarding business insights, context can be seen as information that unites and provides extra dimensions as reference points to more complete evaluation of the situation at hand. If projected against the types of information needs described in Chap. 3, context represents common, or general needs—such information is mostly available all the time, even if important context information over time can get neglected, or new important elements emerge. Context can serve as common denominator—it can be important relevant events or conditions in the background—e.g., dropped oil prices, real estate lending boom, a wave of mergers and acquisitions, etc. Context is intended to add completeness to any important information or insight; however, information integration can overload or distort context, like any other entity of integrated meanings.

Information integration can strip away some richness of the original information, if part of the original context gets lost—fields that were left out; "deep heuristics" that stay with people owning the original data. It leads to a tradeoff between clarity and complexity of the integrated entity—preventing the integration result from being oversized, and preventing loss of important parts of individual context in the process. One of the possible ways to overcome this tradeoff is the involvement of people from fields whose data and information are to be integrated for evaluation of possible loss of context.

Context is inseparable from principal information on a business concern and should be treated as its extension to deliver key points on the background. Often, "full context" or "complete context" are mentioned as a goal; probably the term "sufficient" would be more realistic as a collection of information adequately defining a problem and its environment. Several bare-bone facts on a problem will state the very heart of the matter, but added information on key conditions (When? Where? Who are the actors? Market volatility? Macroeconomic situations? etc.) will expand the picture until it is sufficiently complete. According to Croskerry (2008), context can roughly be split into critical signals (information) and noise. Signals (important new data) arrive with noise; meaningful cues mingle with distracting cues. The value of "sufficient" context is completely arbitrary, but in many areas experience has developed over time that determines the sufficient level of analysis; an example is due diligence procedures when performing the complicated transactions of mergers, acquisitions, market entries and the like, and for a novel situation the definition of sufficient context will be more complicated.

Fischer (2012) separated context elements by their level of definition and structure. Some aspects of context are easy to capture, especially in electronic communications: who—data from login procedure; where—location of access device; when—if a time stamp is provided. In human settings, context parameters are vague and difficult to capture: *concepts* held by individual people; *intentions* that guide their activities; *social relationships* and local cultures that determine interests and behaviors; *values* that underline all of the above.

When integrating data, data elements contained in the records provide rudimentary yet adequate context. When integrating information, the initial snippets of information have their own context that might get lost at the moment of recording or in transmission, so that the bare message is missing some of its key traits; this is one more reason to leave the "last mile" to human judgment.

Wurzer and Smolnik (2008) relate the information integration axis and its specific contexts:

> Given that the user is interested in the business dealings, contracts, products, employees, and service calls of a specific customer, all of this information should be associated with that customer. The information must come from various systems, and should not be fed into a central relational database, in which the linkages would be unalterable. Any possible context is conceivable: A person, a project, a very special document, a meeting, a job, a message, and much more. In each case, the proposed technology framework displays the information containers that are important to that context. A profile of a person may refer to publications within and outside a company, to projects in which that person participated, or to contacts. A document can refer to the author, to other articles enlarging the topic, or to topic areas that are, for example, defined as part of a company's research and development. This results in a new procedure for the retrieval and interpretation of information containers. Instead of searching for terms, or following a directory path, the search focuses on the context that is currently of interest. Naturally, it is also possible to combine contexts to filter information. The user will then immediately have information that is meaningful to a specific context. This allows the user can discover connections in the data inventory and draw conclusions that may result in new knowledge. Users may, for example, recognize that a certain employee has specific skills, which distinguish him in his project experiences, publications, and personal contacts, as well as training.

Data on context from decision support research done in 2004–2007 (Skyrius, 2008) reflect some dynamics of the context—decision information needs known beforehand, including needs on general context, to validate context and make it more reliable; and needs unknown beforehand.

Have There Been Cases when Information Needs Have Been Known Beforehand? If So, What Kind of Information?

The responses to this open-ended question have been grouped into the following groups (Table 5.2).

Table 5.2 Information needs known beforehand (Skyrius, 2008)

Market information—customers, sales, needs, opportunities	49	31%
Competition information—competitors' status, strength, intentions, actions	29	18%
Internal information—financials, capacity, inventory	27	17%
Legal information—laws, regulations, standards	26	16%
No such cases	26	16%
Technical information	2	1%
Total:	**159**	

This group mostly includes information whose content and location are well known and generally accessible because of earlier experience in related cases. It is usually easy to plan and prepare, and this information or its access points can be contained in close proximity to the decision makers.

Have there been Cases when Information Needs have not been Known Beforehand, Having Emerged Only While Making a Decision or too Late Altogether? If So, What Kind of Information?

The responses to this open-ended question have been grouped into the following groups (Table 5.3).

This group of responses can be explained by novelty of the problem or changing conditions. It is hard to plan and cannot be kept prepared and handy; instead, some generic information sources that are most often used in such situations can be made ready to use whenever required, including electronic public archives, search engines, directories etc. A decision support environment providing a set of such sources should be quite helpful. As this activity requires more creative involvement from the users side, a side product of satisfaction of such needs is intensive learning and gained new experience for decision makers.

5.4.3 IT Tools and Methods

The technologies for information integration belong to the advanced segment of informing activities and, according to Levy (1998), are at the intersection of database systems and artificial intelligence. This group of technologies is not so much IT-driven—it tends to lean more towards advanced information management. Probably the most prominent is the area of information extraction tools that examine unstructured data for elements of structure that can be organized and managed. Several examples from published research:

Table 5.3 Information needs not known beforehand (Skyrius, 2008)

No such cases	86	38%
Yes, there have (without specifying the information)	46	20%
Market information	23	10%
Internal information	15	7%
Competition information	14	6%
Legal information	14	6%
Technical information	14	6%
Informal, "soft" information—opinions, foresights	12	5%
Confidential information—e.g., reliability checks	5	2%
Total:	**229**	

- Doan et al. (2009)—a technique for processing unstructured documents by recognizing search arguments in specific positions (title, abstract) and creating structures that can be managed by relational database management system;
- Bernstein and Haas (2008)—extraction of key information from email messages;
- Cody et al. (2002)—a tool called eClassifier to create and maintain a taxonomy from a set of documents;
- Loshin (2013)—pattern-based analysis tools for unstructured text with meta-tagging (tagging in relation to originating context or provenance) and producing probabilities of assigned meaning;
- Ananthanarayan et al. (2013)—a notion of *signature* that summarizes a content of data set for further evaluation of similarity to other data sets, using IBM's LanguageWare tool.
- Holzinger et al. (2013)—a suggestion to enhance human intelligence by computational intelligence, using IBM Content Analytics to enable users to find and recognize previously unknown and potentially useful information.

The extraction of detected elements is usually followed by encoding, clustering and classification procedures.

- **Encoding** applies some set of standards to cover the variety of instances and to assign codes as a token of recognized meaning.
- **Clustering** does group together extracted elements based on assumptions about their meaning. The examples of clusters may be research papers on a given topic; cases of exceptional performance; customer inquiries by topic and channel; customer complaints; quality breaches; failed projects; customer pain-point (a problem that the customer has chosen you to help solving) and many others.
- **Classification** usually is a next step after clustering and aims to arrange the clusters into a structure that reflects a certain logic—hierarchy, consistency, coverage or other.
- An **ontology** can be derived out of this structure, if the subject area and its terms are unambiguous and well-defined. Boury-Brissset (2013) has presented several functions of ontologies regarding information integration:

  - Support standard vocabulary, a taxonomy of concepts and facilitate information sharing;
  - Resolve semantic conflicts;
  - Support text analytics, e.g. Feature, Event, Actor, Information
  - provide semantic annotation;
  - Perform automatic reasoning with business rules.

This sequence is supported in (Schmelzer, 2003) when discussing the process of semantic integration; according to the presented IntegrationZipper approach, the basics of semantic integration are data transformation, data classification into categories, and encoding of unstructured data with metadata.

Shroff et al. (2011) describe a concept of enterprise information fusion, joining internal (mostly structured) and external (mostly unstructured) sources to discover possible risks, and resembling syndication in terms of approach and technology. The concept uses parallel terraced scan—a multi-agented search technique; agents explore many paths but select only the most promising by pruning. The suggested framework monitors external information flows such as news feed or customer feedback, discovers events and fuses this information with internal information to evaluate impact of these events, also covering checks for geographic relevance, supplier impact, and potential threat. Other components are relevant information extraction from internal data, open information extraction from the Web, and Twitter-based opinion mining leading to customer pain point detection.

All the above approaches admit their limitations, largely due to the erratic and incomplete nature of the source information. Cohen (in Hearst et al., 1998), while presenting a WHIRL system to integrate information from the Web, pointed out some limiting issues: the knowledge needed for information integration will be incomplete; integration inferences will be incomplete and uncertain; and mistakes will be made. Schmelzer (2003) admits there are significant barriers for semantic integration. However, recent advances in NLP raise expectations to further push the frontiers of handling sense-making; there also seems to be a growing research interest towards assistance for heuristic information integration, joining together human and IT capabilities.

5.4.4 Heuristic Integration: Dataspaces

In support of heuristic information integration, a number of approaches has been proposed—data spaces and data lakes for storing information without formally relating it to other information; dashboards and mashups to present seemingly related data in a single view; information syndication platforms (RSS) and others.

The approach of *data spaces*, proposed by Franklin et al. (2005), is more of a data and information co-existence approach. The proposed data space support platform (DSSP) must encompass a variety of formats, does not necessarily produce exact results, and has to have tighter data integration tools if necessary. Such features relate data space approach to the idea of tiered information environment, described in Chap. 4. Based on bringing up the "last mile", or human judgment in the end of information integration chain, the data space approach places itself in the group of hybrid intelligence methods.

Dataspace approach is supported by Boury-Brissset (2013), who has suggested a dataspace structure of several layers:

- Segment 0—external data sources and systems containing relevant data;
- Segment 1 (unstructured data)—data store for artefacts;
- Segment 2 (structured data)—data stored using an unified representation scheme;
- Segment 3 (data models)—data models and ontologies to map and integrate heterogeneous data.

Data lakes (Llave, 2018; Miloslavskaya & Tolstoy, 2016) have much in common with data spaces—they also store data and information in their native formats, regardless of structured or unstructured nature. Data lakes are discussed in some more detail in Chap. 7.

Information portals, dashboards and mashups represent an information presentation mode where data and information from different sources are presented together, assembled on user's request (Fig. 5.2). Using portals, dashboards and mashups users may create a context around important indicators, provided that information assembled from various sources is trusted and of compatible quality.

In practical dashboards, there are one or two driving indicators that reflect an important aspect of activities, and other complementary indicators around them, that reflect the context of important aspects or trends. For a certain organization, there are specific key indicators, e.g., occupancy rate; client dynamics; order portfolio and its dynamics; production and inventory, etc. These indicators can serve as an axis for a set of monitoring information watched on a permanent basis. Based on this axis, supporting data can be selected for the informing environment that are easily combined with the axis data (single important axis or several axes representing important activity aspects). This simple information display mode may serve the timely discovery of issues requiring attention.

The selection of information to be monitored on a dashboard raises an interesting issue: what data or information should be monitored together? Just like retail analytics perform a now-ubiquitous basket analysis of goods purchased together, analysis of *"information basket"* (what information should be watched together) may lead to more effective monitoring modes. A simple set of rules be developed on a basis of some standard relations in pairs or triples of information—like "if indicator A goes down together with indicator B, should I check information X?" In terms of research, the existence of information basket may be tested by suggesting a set of monitored indicators and then seeing if removal or absence of one or few of them creates "blind spots", significantly reducing context. Information basket analysis relates to the concept of dataspaces proposed by Franklin et al. (2005); it provides "soft" integration by putting relevant information together, preferably on the same scale if it applies.

A part of research, presented earlier in this book (Skyrius & Bujauskas, 2010), has been directed to the issues of the structure of monitored information by asking the respondents to indicate several most important types of permanently monitored information. Such "basket analysis" of monitored information may provide partial insights on the required structure of the monitored information set. A modest-sized survey (203 filled questionnaires) did not allow for any more significant statistical evaluations; therefore, the analysis just counted the cases of repeating combinations of monitored information. Out of monitored information groups, just the first six groups have been evaluated as being more likely to be included in the sets of information monitored together. Table 5.4 presents the most frequent pairwise combinations of information being monitored together, and Table 5.5 contains most frequently encountered combinations of three.

The most frequent pairs in Table 5.4 point to the closely related information that is relevant to any business activities. The triples in Table 5.5 provide some more

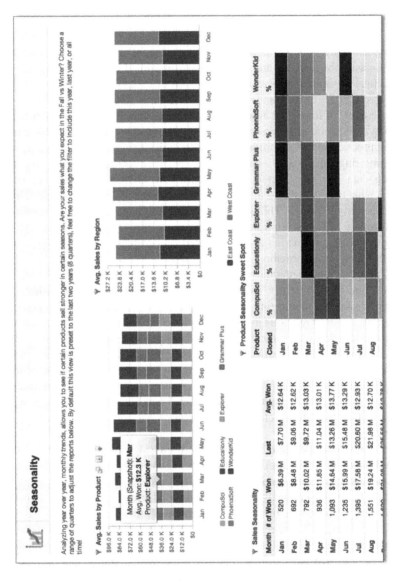

Fig. 5.2 Example of data dashboard (Projects and Dashboards, n.d.)

Table 5.4 Most frequent pairs of monitored information (Skyrius & Bujauskas, 2010)

Information groups		No. of cases
Market and competition	Accounting and financial	16
Market and competition	Sales and turnover	8
Accounting and financial	Sales and turnover	7
Sales and turnover	Customers	5

Table 5.5 Most frequent triples of monitored information (Skyrius & Bujauskas, 2010)

Information groups			No. of cases
Market and competition	Accounting and financial	Customers	10
Accounting and financial	Sales and turnover	Suppliers and inventory	7
Accounting and financial	Customers	Legal regulations	5
Market and competition	Accounting and financial	Sales and turnover	5

information on the mode of environment scanning: groups 1 and 4 aim at a better understanding of market dynamics; group 2 indicates the level of stability of own activities; group 3 concentrates more on the compliance to legal acts. Of course, such simple grouping does not have any practical significance, but further research should increase the scope, significance and reliability of results.

A more substantial analysis of user demand for certain combinations of monitored information might allow inferring about a composite monitoring view where different types of data complement each other. Such "basket analysis" might provide support for satisfying random complex information needs that emerge unexpectedly, by:

- Constant monitoring, or common information needs: designing user's workspace with the sets of key monitored information in mind, and other groups of information can be included in the view if required;
- Supporting discovery—providing possible raw material and tools (easy graphing, combined time series, correlations) for detection of important issues;
- Providing information integration by producing derivative estimates and ratios that allow more everyday insight and diagnostic power;
- Expanding context—triggering additional information needs to elaborate on an issue once it is noticed—assessing payoff/loss, risks, resources available and required, etc.; in general—invoking the second tier of the informing environment with more specialized methods and tools.

Concluding the chapter, we can summarize the most important issues. Data integration, being of technical nature, is structured, deals with data management issues, and sensemaking is mostly left to the users. Standards and data quality are of prime importance.

Information integration deals with all kinds of structured and unstructured data and information to assist sensemaking. The very process and goals of sensemaking are not well defined. While some researchers see information integration and sensemaking as a process with steps and sequence, other researchers stress the non-linear, heuristic and random nature of sensemaking. The definitions of goals range from any advance towards sensemaking to full understanding of the situation at hand. Having in mind that sensemaking is never complete, the goal of information integration and sensemaking may be carefully expressed as a significant advance in understanding an object of interest; the process contains steps but the discovery itself is largely heuristic.

Among the information integration approaches freeform presentation for heuristic integration is an approach that is expected to work when automatic or IT support procedures prove inadequate. This last step is heavily dependent upon wide-scale sense making that, according to Weick et al. (2005), is "social and systemic; uses presumptions and testing; and is about organizing through communication". Many sources agree about leaving the "last mile" to user judgment, where the insight is being finalized.

References

Amaresan, S. (2019). *Data silos: What they are and how to get rid of them.* Retrieved June 20, 2019, from https://blog.hubspot.com/service/data-silos.

Ananthanarayan, R., Balakrishnan, S., Reinwald, B., & Yee, Y. (2013). *Unstructured information integration through data-driven similarity discovery.* New Delhi: IBM India Research Lab.

Andriole, S. (2006). The collaborate/integrate business technology strategy. *Communications of the ACM, 49*(5), 85–90.

Barrett, C. L., Eubank, S., Marathe, A., Marathe, M., Pan, Z., & Swarup, S. (2011). Information integration to support model-based policy informatics. *The Innovation Journal, 16*(1), v16i1a2.

Bernstein, P. A., & Haas, L. M. (2008). Information integration in the enterprise. *Communications of the ACM, 51*(9), 72–79.

Bilhorn, B., & Hulsebus, P. (2014). *10 Best practices for integrating your customer data! Scribe Software White Paper.* Retrieved June 20, 2019, from https://www.scribd.com/document/232426228/10-Best-Practices-for-Integrating-Your-Customer-Data.

Boury-Brisset, A.-C. (2013). Managing semantic big data for intelligence. In K. Laskey, I. Emmons, & P. da Costa (Eds.), *Proceedings of the eighth conference on semantic technologies for intelligence, defense, and security* (pp. 41–47). Fairfax VA, USA, November 12 15, 2013.

Cody, W. F., Kreulen, J. T., Krishna, V., & Spangler, W. S. (2002). The integration of business intelligence and November knowledge management. *IBM Systems Journal, 41*(4), 697–713.

Croskerry, P. (2008). Context is everything, or how could I have been so stupid? *Healthcare Quarterly, 12,* 171–177.

Dervin, B. (1998). Sense-making theory and practice: An overview of user interests in knowledge seeking and use. *Journal of Knowledge Management, 2*(2), 36–46.

Desouza, K. C., & Hensgen, T. (2005). *Managing information in complex organizations.* Armonk, NY: M.E. Sharpe.

Doan, A. H., Naughton, J. F., Ramakrishnan, R., Baid, A., Chai, X., Chen, F., Chen, T., Chu, E., DeRose, P., Gao, B., Gokhale, C., Huang, J., Shen, W., & Vuong, B.-Q. (2009). The case for a structured approach to managing unstructured data. In *CIDR-09. 4th Biennial Conference on Innovative Data Systems Research (CIDR). January 4-7, 2009.* California, USA: Asilomar.

Fahey, L., & Prusak, L. (1998). The eleven deadliest sins of knowledge management. *California Management Review, 40*(3), 265–276.

Fischer, G. (2012). Context-aware systems: The 'right' information, at the 'right' time, in the 'right' place, in the 'right' way, to the 'right' person. *Proceedings of 2012 AVI conference*, Capri, Italy. pp. 287-294.

Franklin, M., Halevy, A., & Maier, D. (2005). From databases to dataspaces: A new abstraction for information management. *SIGMOD Record, 34*(4), 27–33.

Grealou, L. (2016). *Single source of truth versus single version of truth*. Retrieved June 20, 2019, from https://www.linkedin.com/pulse/single-source-truth-vs-version-lionel-grealou/.

Grimes, S. (2008). *Unstructured data and the 80 percent rule*. Retrieved June 15, 2019, from http://breakthroughanalysis.com/2008/08/01/unstructured-data-and-the-80-percent-rule/.

Haas, L. (n.d.). *Information integration isn't simple*. Retrieved June 20, 2019, from http://db.cis.upenn.edu/iiworkshop/postworkshop/slides/Session2Technology1/Haas.pdf.

Halevy, A., Ashish, N., Bitton, D., Carey, M., Draper, D., Pollock, J., Rosenthal, A., & Sikka, V. (2005). Enterprise information integration: Successes, challenges, and controversies. SIGMOD 2005.

Halevy, A., Norvig, P., & Pereira, F. (2009). The unreasonable effectiveness of data. *IEEE Intelligent Systems*, March–April 2009, pp. 8–12.

Hayler, A. (2004). *EII: Dead on arrival*. Retrieved June 21, 2019, from www.intelligententerprise.com/print_article.jhtml?articleID=23901932.

Hearst, M., Levy, A. Y., Knoblock, C., Minton, S., & Cohen, W. (1998). Information integration. *IEEE Intelligent Systems, Trends & Controversies Feature, 13*(5), 12–24.

Holman, V. (2013). *The data integration lifecycle*. Retrieved June 20, 2019, from http://www.victorholman.com/2013/11/07/the-data-integration-lifecycle/.

Holzinger, A., Stocker, Ch., Ofner, B., Prohaska, G., Brabenetz, A., & Hofmann-Wellenhof, R. (2013). Combining HCI, natural language processing, and knowledge discovery – Potential of IBM content analytics as an assistive technology in the biomedical field. In *HCI-KDD Knowledge Discovery from Big Data* (1 ed., pp. 13–24).

Huret, A. (2018). Seeing beyond the big (data) picture. BearingPoint Institute Report. *Interactive*. Retrieved June 15, 2019, from https://www.bearingpoint.com/files/BEI002-Hypercube_seeing-beyond-the-big-data-picture.pdf?download=0&itemId=388636.

Hutton, R., Klein, G., & Wiggins, S. (2008). *Designing for sensemaking: A macrocognitive approach* (Vol. 6). Florence, IT: CHI, 2008 Sensemaking Workshop.

Jennings, P. D., & Greenwood, R. (2003). *Constructing the iron cage: Institutional theory and enactment. R.*

Klein, G. (2013). *Seeing what others don't*. New York, NY: Public Affairs.

Klein, G. (2015). Reflections on applications of naturalistic decision making. *Journal of Occupational and Organizational Psychology, 88*, 382–386.

Kohavi, R., Rothleder, N. J., & Simoudis, E. (2002). Emerging trends in business analytics. *Communications of the ACM, 45*(8), 45–48.

Lau, L., Dimitrova, V., Yang-Turner, F., & Tzagarakis, M. (2014). Understanding collaborative sensemaking behavior using semantic types in interaction data. *Frontiers in Artificial Intelligence and Applications, 2014*, 190–199.

Lee, C., & Abrams, S. (2008). Group sensemaking. *Proceedings of CHI 2008—CHI Conference on Human Factors in Computing Systems*, April 5–10, Florence, Italy.

Levy, A. Y. (1998). The information manifold approach to data integration. *IEEE Intelligent Systems, 1988*, 12–16.

Llave, M. R. (2018). Data lakes in business intelligence: Reporting from the trenches. *Procedia Computer Science, 138*, 516–524.

Loshin, D. (2013). *TDWI checklist report: Integrating structured and unstructured data. TDWI Research.*

Madsbjerg, C. (2017). *Sensemaking: The power of the humanities in the age of the alghorithm.* London, UK: Little, Brown.

Miller, J. (2000). *Millenium intelligence. Understanding and conducting competitive intelligence in the digital age*. Medford, NJ: Cyber Age Books.

Miloslavskaya, N.G., & Tolstoy, A. (2016). Application of big data, fast data and data lake concepts to information security Issues. *The 3rd International Symposium on Big Data Research and Innovation* (BigR&I 2016).

Mudrik, L., Faivre, N., & Koch, C. (2014). Information integration without awareness. *Trends in Cognitive Science, 18*(9), 488–496.

Park, S.-H., Huh, S.-Y., Oh, W., & Han, S. P. (2012). A social network-based inference model for validating customer profile data. *MIS Quarterly, 36*(4), 1217–1237.

Pirolli, P., & Card, S. (2005). The sensemaking process and leverage points for analyst technology as identified through cognitive task analysis. *Proceedings of International Conference on Intelligence Analysis*, May 2005, McLean, VA.

Portela, F., Aguiar, J., Santos, M. F., Silva, A., & Rua, F. (2013). Pervasive intelligent decision support system—technology acceptance in intensive care units. In A. Rocha et al. (Eds.), *Advances in Information Systems and Technologies, AISC 206* (pp. 279–292). New York: Springer.

Projects and Dashboards and Tabs. (n.d.). Retrieved June 20, 2019, from https://help.gooddata.com/doc/en/reporting-and-dashboards/dashboards/using-dashboards/projects-and-dashboards-and-tabs.

Redman, T. (2004). Data: An unfolding quality disaster. Accessed January 15, 2006, from www.information-management.com/issues/20040801/1007211-1.html.

Roth, M. A., Wolfson, D. C., Kleewein, J. C., & Nelin, C. J. (2002). Information integration: A new generation of information technology. *IBM Systems Journal, 41*(4), 563–577.

Russom, Ph. (2007). *BI search and text analytics. New additions to the technology stack. TDWI Best Practices report.* Retrieved June 25, 2019, from http://download.101com.com/pub/tdwi/Files/TDWI_RRQ207_lo.pdf.

Salmen, D., Malyuta, T., Hansen, A., Cronen, S., & Smith, B. (2011). Integration of intelligence data through semantic enhancement. Proceedings of the 6th International Conference on Semantic Technologies for Intelligence, Defense, and Security (STIDS 2011), George Mason University, Fairfax, VA, 6-13.

Saunders, C., & Jones, J. W. (1990). Temporal sequences in information acquisition for decision making: A focus on source and medium. *The Academy of Management Review, 15*(1), 29–46.

Schmelzer, R. (2003). *Semantic integration: Loosely coupling the meaning of data.* Retrieved June 20, 2019, from https://doveltech.com/innovation/semantic-integration-loosely-coupling-the-meaning-of-data/.

Schneider, M. V., & Jimenez, R. C. (2012). Teaching the fundamentals of biological data integration using classroom games. *PLoS Computational Biology, 8*(12). Retrieved May 3, 2018, from https://journals.plos.org/ploscompbiol/article?id=10.1371/journal.pcbi.1002789.

Shroff, G., Aggarwal, P., & Dey, L. (2011). Enterprise information fusion for real-time business intelligence. *14th International Conference on Information Fusion.* Chicago, Illinois, USA, July 5–8, 2011.

Skyrius, R. (2008). The current state of decision support in Lithuanian business. *Information Research, 13*(2), 345. Retrieved March 1, 2018, from http://InformationR.net/ir/31-2/paper345.html.

Skyrius, R., & Bujauskas, V. (2010). A study on complex information needs. *Informing Science: The International Journal of an Emeerging Transdiscipline, 13*, 1–13.

Skyrius, R., Šimkonis, S., & Sirtautas, I. (2014). Information integration: Needs and challenges. *Information Sciences. Research papers. Vilnius University, 69*, 74–88.

Verhoef, P. C., Kooge, E., & Walk, N. (2016). *Creating value with big data analytics.* Abingdon, UK: Routledge. Retrieved June 25, 2019, from https://www.inetsoft.com/business/solutions/what_is_data_mashup_what_are_benefits/.

Wang, R. Y., & Strong, D. M. (1996). Beyond accuracy: What data quality means to data consumers. *Journal of Management Information Systems, 12*(4), 5–33.

Weick, K. E. (1995). *Sensemaking in organizations.* Thousand Oaks, CA: Sage Publications Inc..

Weick, K. E., Sutcliffe, K. M., & Obstfeld, D. (2005). Organizing and the process of sensemaking. *Organization Science, 16*(4), 409–421.

White, M. (2018). *80% of corporate information is unstructured. Really?* Retrieved June 25, 2019, from http://intranetfocus.com/80-of-corporate-information-is-unstructured-really/.

Williams, S. (2016). *Business intelligence strategy and big data analytics. A general management perspective*. Cambridge, MA: Morgan Kaufmann.

Wurzer, J., & Smolnik, S. (2008). Towards an automatic semantic integration of information. In L. Maicher & L. Garshol (Eds.), *Subject-centric computing. Fourth International Conference on Topic Maps Research and Applications* (pp. 169–179). Leipzig, Germany: TMRA.

Chapter 6
Management of Experience and Lessons Learned

6.1 Terms and Definitions

Accumulated expertise and its reuse is one of the principal value drivers in any activities in current competence- and innovation-rich economy. As any other useful information, accumulated expertise has a significant role in BI as tested facts and conclusions about situations that actually have happened, have a degree of similarity to the problem at hand, and whose outcomes are known. Cyert and March (1963), while discussing a related area—organizational learning, have argued that a firm learning from its experience leverages rules, routines and procedures intended to respond to external shocks, and by doing this strengthens organization agility and adaption to the environment changes. Likierman (2020), discussing good judgment in making top-level decisions, stressed the importance of experience as a source of context and objectivity. By leveraging expected repetition of situations and solutions, preserved expertise supports avoiding "invention of bicycle". The organization's information systems keep records of transactions; however, data in transaction databases do not provide adequate coverage of important situations and their outcomes. In its own turn, recorded problem-solving expertise covers (or is expected to) more information about actual situations and problems. As well, this information covers only part of the value-creating expertise; nevertheless, the capture and record of this expertise might assist in answering the following questions:

- Roots of problems that have happened;
- Problems that could have been avoided;
- Problems that were avoided and are to be avoided in the future;
- Mistakes not to repeat in the future;
- Opportunities that were not used;
- The right steps that had been taken, or things that have been done well;
- Possible improvements;

© Springer Nature Switzerland AG 2021
R. Skyrius, *Business Intelligence*, Progress in IS,
https://doi.org/10.1007/978-3-030-67032-0_6

Records of complex transactions and experiences can produce ground for rich analytics that support extraction of important derivative information:

– Similar features and their measurability;
– Level of similarity—indexes, coefficients, weights etc.;
– Context definition and comparison issues;
– Evaluation criteria and their changes over time etc.

Actually, using or analyzing experiences is a research task—e.g., "show all understaffed projects" or "find out how decision scenarios have been developed". The above arguments relate to cases that are ideal or close to ideal—there's a specific information need; and there's a collection of specific information to cover that need. A practical controversy that emerges quite often, however, is between an acute need for specific earlier experience in difficult situations, and reluctance to record and preserve this experience.

Many rules apply to experience information in similar way as to BI main product—current intelligence, or current insights. Collecting, organizing, using, improving collection and use skills—these are the features specific to both current BI and LLs. The main differences are the latency of LLs, as well as the trust in LL as time-tested information. The important role of LL in training and developing BI competencies rests on real-life cases with known outcomes.

O'Donnell, Sipsma, and Watt (2012), while discussing the critical issues facing business intelligence practitioners, have stressed the importance of preserved expertise in motivating the future users of BI systems:

> Engaging business users was considered to be a significant challenge, especially early in a new BI system development project. The practitioners were very interested in stories, case studies and best practices to get business users to value the BI systems that were developed for them. It is hard to get end users to develop energy and excitement for the change the systems potentially can enable. Ongoing training and education was seen as a useful tool not only to teach end users about systems capabilities but to interest them in what the systems could mean for their business. The training of power users and their ability to act as evangelists for the uptake of systems was seen as very important. Further, *as people move around and out of an organization it is important that the knowledge of the value of the BI systems they have gained isn't lost* and, in the case of power users, the energy and drive to use the systems they provided needs to be replaced.

The availability or lack of experience information, as well as willingness or reluctance to share it leads to important cultural issues that are discussed in more detail in Chap. 9.

6.1.1 Terms

The strive for effective organization and reuse of valuable experience has initiated many research directions, and a set of specific related terms has emerged over time: *experience base, expertise management, lessons learned, knowledge management, communities of practice, best practices, organizational learning, organizational memory* and others, knowledge management being the most prominent.

One of the early approaches to manage expertise has been knowledge management (KM), based on information technology, and with rather disappointing results. Fleisher (2003), while discussing the issues of knowledge management, indicated different activities that could potentially fall under the KM umbrella: knowledge-mapping, data- or knowledge-mining, knowledge audits, knowledge databases, corporate intranets or digital library development and maintenance, personal and virtual navigation, corporate knowledge directories, FAQ development and so on. Alvesson (2004) has pointed out that often problems of LL are pushed into knowledge management framework, although IT-based knowledge management practices have often shown that more management means less knowledge.

Easterby-Smith and Lyles (2011) have discussed a variety of terms from the above-mentioned sets, namely, organizational learning, learning organization, organizational knowledge and knowledge management. Using a distinction between the first two terms articulated by Tsang (1997), they stated that *organizational learning* is more process-oriented, while the *learning organization* features effective learning as a precondition to prosperity. Regarding *organizational knowledge* and *knowledge management*, they noted that many sources, discussing the former, adopt a philosophical attitude towards definition of knowledge in organizations, while the latter is often discussed in terms of technology platforms and procedures of "measuring, disseminating, storing, and leveraging knowledge in order to enhance organizational performance".

Etienne Wenger (2006) has pointed to **communities of practice** (CoP) as an alternative approach to manage expertise. CoP cover domain, community and practice, and focus on people, as opposed to focus on content in knowledge management. CoP establish a direct link between learning and performance, attempting to overcome the inertia of information "stickiness", discussed further in this chapter. They are not limited by formal structures, therefore challenges for the traditional hierarchical organization arise.

There are no clear borders for all terms mentioned above in this paragraph. For these terms, the uniting feature is the drive to capture, preserve and leverage valuable experience as information (labelled knowledge in some sources) that can be a powerful asset. The creation and use of this asset can be supported by technology tools like creation and use of any other information.

The difference between experience and lessons learned seems rather thin; yet some sources provide arguments to distinguish the two. Weber, Aha, and Becerra-Fernandez (2001) point out that experience is potentially valuable information that has been captured, collected, stored, distributed and possibly reused, and can be seen as the "lessons" part without being learned. Lessons learned are often defined differently—as significant change resulting from utilizing a lesson (Weber et al., 2001). One of the most complete definitions is provided by (Secchi, Ciaschi, & Spence, 1999):

> A lesson learned is a knowledge or understanding gained by, or a derivative of experience. The experience may be positive, as in successful tees or mission, or negative, as in a mishap or failure. Successes are also considered sources of lessons learned. A lesson must be significant in that it has a real or assumed impact on operations; valid in that it is factually and technically correct; and applicable in that it identifies a specific design, process, or decision that reduces or eliminates the potential for failures and mishaps, or reinforces a positive result.

To avoid confusion, a term "lessons learned" (abbreviated LL) will be used further in this book.

6.1.2 Positioning

One of the possible ways of positioning the terms and environments related to LL is presented in the Fig. 6.1 below.

Several examples to illustrate the relation in Fig. 6.1 are presented below in Table 6.1.

Not all gained LL are recorded, not all recorded LL fall neatly into predefined structures, not all recorded LL are used, and not all LL are recognized as useful or relevant, as shown in the Fig. 6.2.

The above separation of LL into recorded and unrecorded, structured and unstructured, used and unused does not mean that only the useful and relevant LL should be given attention and transferred into records. It is difficult, or often impossible, to

Fig. 6.1 *Relationship between terms and environments*

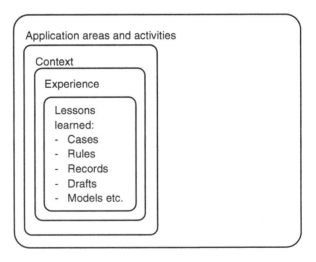

Table 6.1 Examples of relations between application area, context, experience and lessons learned

Application area	Context	Experience	Lessons learned
Market strategy	Global economic outlook; features of a certain export market	The availability of a local partner saves time in export market entry	When entering a new export market, a safe approach is to have a local partner or to acquire one
Business intelligence	Expansion of BI user base; need to democratize reporting	New users have motivation but lack skills	Expand report creation capability by training users
General strategy	Ethical norms and climate	Neglect of ethical norms and values in the past has created serious image problems	Always evaluate corporate social responsibility issues when defining business strategy
Process management; contingency planning	Turbulent environment; possible unexpected problems	A strained process load makes the process and its connections fragile	Never exceed 80 percent of process capacity

Fig. 6.2 Positioning of recorded LL

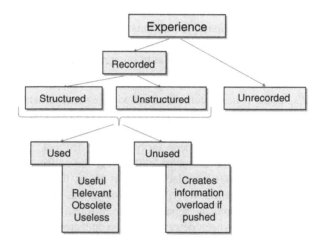

guess the future relevance of the current LL, so their preservation may be seen more like a form of insurance against future information needs.

6.1.3   Content

The content of LL is usually intended to contain concise description of decisions, approaches, frameworks, outcomes, explanations and other information that is considered vital to consistent presentation of important issues in the past. According to Patton (2001), high quality lessons learned include:

Table 6.2 Types of captured experience (Weber et al., 2001)

Knowledge artifacts	Originates from experiences?	Describes a complete process?	Describes failures?	Describes successes?	Orientation
Lessons learned	Yes	No	Yes	Yes	Organization
Incident reports	Yes	No	Yes	No	Organization
Alerts	Yes	No	Yes	No	Industry
Corporate memories	Possibly	Possibly	Yes	Yes	Organization
Best practices	Possibly	Yes	No	Yes	Industry

1. Findings—patterns across programs.
2. Basic and applied research.
3. Practice wisdom and experience of practitioners.
4. Experiences reported by program participants/clients/intended beneficiaries.
5. Expert opinion.
6. Cross-disciplinary connections and patterns.
7. Assessment of the importance of LL.
8. Strength of the connection to outcomes attainment.

Weber et al. (2001) have presented several types of captured valuable experience, discussing it in the context of knowledge management (Table 6.2).

We may note that the separation of successes and failures into two different types, while potentially interesting in terms of specific features for positive or negative outcomes, wouldn't necessarily happen in real applications. During research performed by the author in 2008–2009, the collected survey data (collected from 204 respondents) included responses about the content of accumulated decision making experience by types of information. As the subject of the survey were business decisions, no separation into different collections for positive and negative experiences has been detected, and the collected experience information, positive or negative, is collected and kept together. The aggregated survey data, which has not been published before, is presented below in Table 6.3.

The above information groups can be split into information about the problem, and information about the context. The problem itself is described by:

• Decision approaches and methods;
• Short description of the problem;
• Problem size or value;
• Decision alternatives.

The most common problem context features are:

• Time of problem emergence and the decision made;
• Market situation;

Table 6.3 Most common experience information about taken decisions (unpublished research data)

Information types	Count
Decision approaches and methods	178
Short description of the problem	150
Problem size or value	93
Time of problem emergence and decision made	91
Decision alternatives	88
Market situation (own share, maturity, competitiveness)	72
Location where the problem has emerged	71
Decision working matters—drafts, working papers etc.	67

- Location where the problem has emerged.

There appears to be a reasonable set of data reflecting the principal features of the problem or decision—about 5–6 data elements, and another similarly sized set of data reflecting the context. Some of the important features are just numerical values—date, location, problem size; while other features can only be presented in text format—decision approaches and methods, short description of the problem, market situation etc. There hardly is a single type of format for LL records, and the variety of forms is briefly discussed in the following paragraph.

Important features for LL are elements of structure that ensure adequate capture of lesson features for future uses. Some may be obvious and easily coded, like date, transaction volume or location; some may be coded into categories or types (market situation, problem type) with risk to lose richness. There are activities that feature compulsory records of events—healthcare (e.g., surgeries), police work (e.g., road accidents) and others. Such records possess more complex structures and richer content than simple business transactions, and present a valuable source of experience for future practices. Over long-time practices, specific requirements for content of records have been determined that raise no doubt about their accuracy, precision or completeness. This is not so evident for numerous other activities—e.g., business decisions leave a trail of information describing the problem and decision approaches, but the larger the variety of problem situations, the more complicated is the introduction of record standards.

Issues with LL structure and forms emerge and are discussed below.

6.1.4   Structure and Forms

Lessons learned, like other activities named in this paragraph that deal with valuable information, can be seen as an area of advanced information management with its own requirements. Experience is recorded largely in "soft", or unstructured, format; its proper capture requires extra effort while its value is not obvious; its representation is complicated by attempts to introduce structure while maintaining full meaning and context for conveying it for appropriate situations or relating this experience to other experiences. A saved instance of LL is not exactly a database record of business transaction, nor is it a purely text document. The procedures of using LL involve both structured and unstructured data. According to Weber et al. (2001), often lessons are collected in text form, and then supplemented with structured fields (see Chap. 5 on information integration).

The records of LL can take different forms—cases (Weber et al., 2001; Yan, Chaudhry, & Chaudhry, 2003), rules (Weber et al., 2001), lesson reports (Rowe & Sikes, 2006; Seaman, Mendonca, Basili, & Kim, 1999), event descriptions, best practices (Patton, 2001; Weber et al., 2001), alerts, incident reports (Weber et al., 2001), tips (Yamauchi, Whalen, & Bobrow, 2003). The recorded experience information has dimensions of its own: it may range from case-specific to repetitive and general; from formal/structured to informal/unstructured; from useful and relevant to irrelevant or useless and so on.

Against a different perspective, Weber et al. (2001) discuss cases and rules under a joint term of knowledge representation.

Cases are more related to a specific task, compared to LL that intend its records to be reused for a variety of tasks. Yan, Chaudhry, & Chaudhry (2003) present features of case-based reasoning, focusing on reuse of rational experience, case features, properties and representation, characteristic and non-characteristic properties. Case-based reasoning (CBR) is a variation of expert system aimed at solving the problem at hand by using available analogies from the past in the case base. The paper describes the use of CBR in solving third-party logistics evaluation problems by estimating the degree of similarity between past cases and the present problem, using a comprehensive evaluation index system, and an object-oriented approach for case representation.

Rules reflect expressions of logic that has been recognized and tested in real conditions, and, compared to LL, are better defined and therefore easier to reuse. Business rules use more formal representation—systems, standards, languages. However, the inherent formality of rules requires a near-perfect matching of conditions, whereas LL require just a partial matching.

Considering the relation of business rules to BI, business rules are generalizations out of numerous occurrences; this is not always the case with BI where situations may vary from rather standard and therefore rule-friendly to specific or unique. Business rules set a part of business logic, although may have different powers (from legal power to recommendation), and have to allow convenient modification, if they have to be re-tuned to changes in business environment. If easily edited, rules may provide a basis for much-needed business agility. An important feature of rule application is that instead of being sealed inside software code, rule parameters or controls are placed outside code for easy access and editing. Rules define certain (non-fuzzy) logic, while LL can be vague. The typical examples of rules-heavy industries are finance and insurance (Hildreth, 2005). We can note that use of rules for BI should be limited because of focus on operations level.

A couple of examples of forms for capturing LL are in Figures 6.3a, b.

University Services
program.management.office

PROJECT LESSONS LEARNED REPORT

Project Name:	Business Intelligence (BI)
Prepared by:	Diane Kleinman
Date:	June 15, 2009

Project Close-Out Discussion

A Lessons Learned meeting was held on 6/12/09. The summarized lessons learned survey results are attached to this document.

Attendees:	
Vel Angamthu	Janet Heller
Wendy Berkowitz	Bill Kanfield
Wayne Bowker	Ann Lundholm
Aaron Demenge	Ron Mapston
Michael Garza	Peggy McCarthy
Jim Green	Tammy Nelson
George Hardgrove	Bill Paulus
Janet Heller	Jennifer Pierson
Bill Kanfield	Shari Zeise

List this project's biggest successes.

Description	Factors that Promoted this Success
We have a more organized reporting structure.	The capabilities of BI allow for a more organized reporting structure
We were able to remove a lot of reports that weren't being used and we have reports that actually work.	A benefit of this project was time was taken to examine existing reports and to remove those reports that no longer were needed or did not work.
The tool will allow users to write their own reports and these can be modified on an ongoing basis	BI allows more than just a few people to have the capability to create reports. There is no bottleneck like there was before when only one person knew and was able to write reports that are needed. Now users can write their own reports and reports can be modified on an ongoing basis.

List areas of potential improvement along with *high-impact* improvement strategies:

Category	Project Shortcoming	Lesson Learned
Project Management	There are still questions around whether or not to still treat this as a project and let the team make decisions on how to move forward with operational monthly reports.	FM Leadership needs to determine if this project should continue with a project structure in place. There needs to be a focus on completing operational monthly reports.
Project Communication	The right people were not always included on project teams and sometimes the teams needed to change mid-stream as the project requirements changed.	Continually monitor the makeup of project teams to make sure that the right people are included when requirements change.

Fig. 6.3 (**a**) A generic project lessons learned template (Source—FormsBirds, https://www.formsbirds.com/free-project-lessons-learned-report). (**b**) A PRINCE2 template for capturing lessons learned (Source—http://wiki.doing-projects.org/index.php/Lessons_learned_-_a_tool_for_sharing_knowledge_in_project_management)

LESSONS LEARNED LOG						
Project Name:		<optional>				
National Center:		<required>				
Project Manager Name:		<required>				
Project Description:		<required>				
ID	Date Identified	Entered By	Subject	Situation	Recommendations & Comments	Follow-Up Needed?

Fig. 6.3 (continued)

A widely known form of preserved LL is bodies of knowledge (BOKs) and other widely accepted collections of experience and lessons learned. Important traits for BOK, as well as for any collections of experience are structure, contribution and use. For BI, a development of BOK is complicated because of strategic and confidential nature of BI activities.

6.2   Lessons Learned Theoretical Issues

The published research, aimed exactly at LL, experience management or competence preservation under exactly these terms is not very abundant. However, many related areas exist that have attracted research attention and address important features of LL.

In the early stage of knowledge management research wave, Cohen and Levinthal (1990), while focusing on the use of external experience and knowledge, defined organization's **absorptive capacity** as prior related competence that "confers an ability to recognize the value of new information, assimilate it, and apply it to commercial ends". This definition rather well applies to the potential of using LL in BI, providing a concept of measure for LL demand (on the receiving side): the capacity to assimilate and reuse internal experience is as well vital for future value creation as external experiences. Absorptive capacity is a feature of organizational intelligence level, and creates value by using specific, often new information that is perceived as important for the organization.

According to Šimkonis (2016), to qualify as LL, the events in any activities can be evaluated against the dimensions of their importance and perception. Some events are more important than others, and some events are more easily perceived than others; thus the events may be classified into four quadrants, as shown in the Fig. 6.4.

The first quadrant contains events that are unimportant and unperceivable. In the second quadrant, the events are perceivable but unimportant or neutral, and do not deserve time or other resources. The third quadrant contains important events; however, their meaning stays unperceivable to the participants. Regardless of the abundance of events belonging to the first three quadrants, no lessons will be created.

	Unimportant	Important
Unperceivable	1. Unimportant, unperceivable	3. Important, unperceivable
Perceivable	2. Unimportant, perceivable	4. Important, perceivable

Fig. 6.4 The Importance and perceivability of events (Šimkonis, 2016)

For lessons learned the most important one is the fourth quadrant containing important situations that are perceived by the participants—important and perceived situations may be understood as sources of lessons. As well, events in the quadrant of important yet unperceivable are potential source of LLs, once the nature of the events becomes clear over time. As well, unimportant events may appear important over time.

It has to be noted that the definition of the above quadrants is subjective: importance or perception belong upon the reflection skills of the participants. Regardless the imprecise nature of this classification, it assists in defining the field of higher attention—the fourth quadrant containing important and perceivable events. Also, as time goes by, unperceivable important events become clearer and in most cases ultimately perceivable, thus becoming sources of lessons.

Evans (2003) has defined organization community types that influence the formation of expertise:

- Bureaucratic community—the area of "known knowns"; e.g.—accounting systems. Risks emerge in failure to notice changes; best practices often are contained in executives' brains and may become a factor of stagnation.
- Expert community—the area of "known unknowns". Largely based on recognition and analysis; learning is respected; experience rests in experts' minds. Popular among knowledge management proponents.
- Shadow community—the area of "unknown unknowns". Based on a network of informal communities; supports experiments and a right to make mistakes; experience is permanently renewed and accumulated from common experiments and community platforms. Risks emerge in temptations to return to relations of traditional management and control.
- Chaotic community—messy but fruitful for innovations. Deals mostly with unstructured problems; no experience.

The context in which situations happen and expertise develops, as shown in existing research, is of prime importance. As mentioned earlier in the book, Snowden and Boone (2007) have developed a Cynefin model for describing types of contexts

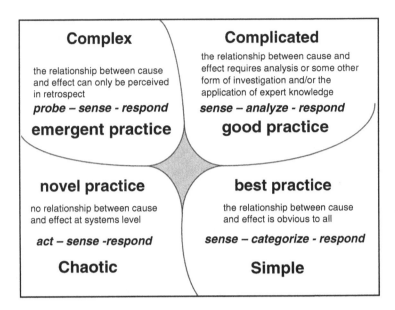

Fig. 6.5 Snowden and Boone's Cynefin model for context types (Snowden & Boone, 2007)

for decision making along the dimension of perceivability, and defined four types of contexts (Fig. 6.5): simple, complicated, complex, and chaotic.

According to the Cynefin model, context depends on the novelty of situation, and relates to Simon's levels of decision structuration. The role of LL changes from best practice to scarce practice with increasing uncertainty (counter-clockwise), or from scarce to best practice with increasing certainty (clockwise). As expertise accumulates, under normal course of events more and more uncertain issues are clarified and become certain. The opposite "jump" from simple to chaotic context is possible in threshold situations caused by bias and ignorance, complacency or neglect. When some expertise is lost (people leave or die), or there's a paradigm change, counter-clockwise context changes occur—the context switches from simple to chaotic, complex, or at least complicated.

We can note that Evans' community types roughly correspond to Snowden and Boone context types; an attempt to project the context and use of experience into Evans' community types has been described in (Šimkonis, 2016):

1. *Simple contexts* and *bureaucratic community*: contexts are simple and stable; the issues to consider are "known knowns"; use of experience or lessons—best practices are widely used; activities—sense, categorize and respond; risks—oversimplification; tunnel vision; failure to notice important changes.
2. *Complicated contexts* and *expert community*: muddled contexts with unclear causes/effects; more than one right solution is possible; the issues are "known unknowns"; good practices instead of best practices; sense, analyze and respond; risks—tunnel vision at expert level, analysis paralysis.

3. *Complex contexts* and *shadow communities*: context is a continuous flow of changes; hard to define right solutions; "unknown unknowns"; probe, sense and respond; informal expert communities; experimentation and tolerance of mistakes; risks—impatience with results may shift back to authoritative management style.
4. *Chaotic contexts* and *chaotic communities*: contexts—turbulence; no right solutions and good practices; act, sense and respond; no good practices or any practices; risks—management based on action may have trouble adapting once the crisis is over.

A dynamic relation between events, context and community types is possible because of changes over time in several aspects: perceivability of events that grows with time; clarification of context over time; changes in community as the context evolves from uncertain to more certain. However, according to Gill's idea of local ruggedness (Gill, 2008), other, new contexts may emerge in the background. If projected against Chris Argyris double loop learning model (Argyris, 1991), changes in context may initiate the move to the second loop. Perceivability of events may change fast, as does the context; community is more inert to changes and migration across types. When the certainty of context grows, community should move from fluid and creative to more institutionalized one. This move may be valid just for some activities that have experienced changes. The changes named in this paragraph may affect such features of LL as validity span ("does not apply anymore"), reliability ("cannot be applied to specific groups of cases") and overall value.

The dynamic dependence of LL upon changes in context and community types suggests relations to BI maturity. In BI maturity models, the initial (1st) stage defines a phase containing no experience and often erratic; the later stages use the accumulated expertise and make the process easier to define and manage. This expertise may have dual influence on BI performance: while some part of it is universal expertise that maintains its significance over time, other expertise might be specific to the period or technology, and lose its value over time. As such expertise becomes obsolete or misleading, a need emerges for expertise review and evaluation that are part of a wider set of BI governance procedures to manage and reuse BI assets.

Discussing dynamic capabilities, Zahra, Sapienza, and Davidsson (2006) have noted: "Managers do not, and probably should not, create 'once-and-for-all' solutions or routines for their operations but continually reconfigure or revise the capabilities they have developed. ... The new routines form the foundation of firms' knowledge bases". However, old routines are hardly discarded; they remain in some form of experience collections, although as valid competences they are no more useful. Having lost their relevance, they, however, form a sort of timeline or time series of competence changes, and may be useful in forecasts for new competences.

Management of LL can be one of the sources of BI agility—people change, as in project management, but lessons remain as a foundation of resilience to withstand changes. There is a risk of stereotypes, but that's a risk featured by all aging information. Having been very true at the moment of its creation, such information—rules, frameworks, heuristics, past projects and decisions—is challenged by changing environment, and some part of this information inevitably becomes obsolete. Issues of BI maturity and agility are discussed in more detail in the Chap. 8 of this book.

6.3   LL Management Process

Records of LL summarize the work done, define its context, evaluate the results, and range in formality from crude drafts to developed structures and collections. LL intertwine with areas of sophisticated information management like decision support and organizational learning, communities of practice, and other forms of capturing valuable information for future use. All the above areas use advanced techniques and sophisticated innovations to manage content; it would likely be logical to assign the area of managing LL to *advanced information management*.

LL management, like any other intelligent process of information management, has a supposedly cyclical structure with feedback (Fig. 6.6). Like in other processes, the process is fed by arrival of new information—in this, case, new experiences that are verified, adjusted to structural requirements and stored. The use of stored LL justifies their value and provides feedback for collection of future LLs.

The dynamics of environment changes seems to have a twofold effect on value of LL. On one hand, more frequent similar situations requiring decisions create additional demand for LL as important source of decision arguments. Any processes of repetitive nature create a body of stable experience, and are fruitful ground for their eventual automation.

Fig. 6.6 The cyclical structure of the process of LL management

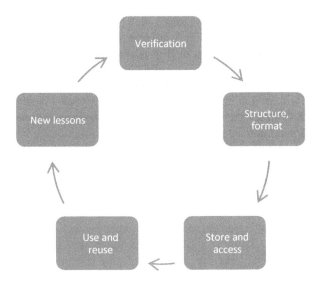

On the other hand, in turbulent environment situations are less likely to repeat themselves, thus diminishing the motivation to use experience information and reducing its value. A specific area with low repetition of situations is project management. As very few projects are alike, the probability that a large part of one project's experience can be reliably reused in another project, is expected to be rather low (Šimkonis, 2016). This is clearly reflected in PRINCE2 definition of the project as "a temporary organization, established for creation of one or several products" (Managing Successful Projects, 2009). We have to note that BI is neither purely functional or project-oriented, and its fluid and constantly changing nature creates a variety of experience that requires a variety specific methods of handling LLs.

Published sources on the topic of LL management accentuate stages of development, dynamics, verification, blend of push and pull approaches, LL analytics for the cases where deeper inquiry in required. Discussing dynamics of LL management, Rowe and Sikes (2006) suggest different levels of lessons learned and advice for transition to higher levels:

Level 1: Lessons learned process (what is required).

No routine capture; no process, tools or techniques in place; no consistency. Lessons handled on *ad hoc* basis. Process should include 5 steps:

- **Identify** stuff valuable for future work—three key questions: what went right, what went wrong and what needs to be improved
- **Document** the lessons—record what has been identified; a danger of over-bureaucratizing;
- **Analyze**—information is discussed and shared between teams; improvements are the usual product of analysis;
- **Store**—*ad hoc*; no standards
- **Retrieve**—complicated for level 1

Level 2: LL management process becomes part of the organizational culture, and acquires consistent processes and forms.

Level 3: Metrics are introduced wherever possible to increase manageability.

The suggested approach is reminiscent of the maturity phases that lead from fluid initial phase to managed and measured processes in the later phases. The risks of not doing the above are rather clear: repetition of mistakes—own or somebody else's; and missed opportunities to leverage existing lessons.

Just like database technology allows extended manipulation of data using OLAP, data mining and other techniques, advanced information management and other aspects of advanced informing are expected to provide a higher-level information integration in collections like expertise base or LL, and perspectives by which different users can access, select and consolidate relevant information fragments in required ways. If the issues of information integration will receive considerable advance, this should be a significant step to an environment where information collections are manipulated with ease comparable to today's data manipulation.

Patton (2001), discussing the institutionalization of LL management, points out that high-quality LL have to be independently triangulated to increase transferability as cumulative expertise that can be adapted and applied to new situations. We have to note that compulsory triangulation of all LL would be one more informing procedure to perform, and seemingly redundant to many people involved. On the other hand, proper structuration of content and context would allow easy integration of LL by providing necessary axes for triangulation and subsequent rise of LL quality. That would be especially obvious when in a problem situation important relevant LLs would be verified before or during problem solving process. Šimkonis (2016) has defined the factors that affect the decision to assign resources for LL analysis:

- Experience on the results of former cases of LL analysis—were they of considerable value, or just a time waste?
- General attitudes towards assigning resources for LL analysis—more support for analysis or for faster action?
- The individual preference to analyze or act.
- The scope of possible consequences.

Some of the above factors are difficult to handle, but their recognition leads from mechanic record of LL towards a wider understanding of a problem and better decision quality.

Other methods and tools exist for effective management of LL—for example, follow-up on recorded LL to estimate whether they have attracted anybody's attention; or discussion thread metrics to measure popularity of stored experience. However, their value and the practical issues of their use are rather under-researched for the time being.

6.4 Practical LL Management Implementation Cases, Systems and Features

There are cases of LL management implementation which, however specific, present arguments about value of LL management and actual systems that create and deliver this value.

According to Weber et al. (2001), the early applications of LL management were implemented in the military domain, NASA and other space agencies. In some cases, their nature is conditional and might resemble the logic of an expert system or business rules: "if condition X is satisfied, lesson (experience) Y is applicable".

In early 1990s Xerox had developed an information repository, named Eureka, for technicians to share their tips for solving tricky equipment problems. Labelled at its launch as a "knowledge management system", in today's context after over 20 years in operation it has gone global and can be considered a company-wide LL system. Yamauchi et al. (2003), analyzing the experience of using Eureka, have noted that for users-technicians the use of information is a mix of formal Repair Analysis Procedures (RAP) and gleaning—random collecting of tips, rumors etc. Solving a certain problem includes using all possible information sources—coworkers, service calls, the log book, system messages, the tips from Eureka system designed to share practical information among technicians, and surrounding environment. Doyle (2016) has pointed out factors for such system to be effective:

- Presence of an organizational culture supporting promotion and sharing of knowledge;
- Easy access to information;
- Adequate staffing to ensure quality of data;
- Buildup of trust in information by having expert technicians vet all tips submitted to the platform.

Miller (2000: 138), while discussing intelligence archives, pointed out some important features of a good archive system: full-text search for information retrieval; provenance information including author-stamp and date of storage; avoiding duplication of commercial information systems. He also noted that existing technologies for storing and exchanging are simple, but many companies have difficulties finding time for their maintenance.

Saran (2019) has described a story of how an idea of accumulating recorded scientific expertise turned into something global—no less than the World Wide Web. Thirty years ago, while working at CERN, Tim Berners-Lee stumbled upon a problem of maintaining valuable scientific information for potential use by fast-changing body of researchers.

"A problem, however, is the high turnover of people. When 2 years is a typical length of stay, information is constantly being lost. The introduction of the new people demands a fair amount of their time, and that of others, before they have any idea of what goes on," he wrote. "The technical details of past projects are sometimes lost forever, or only recovered after a detective investigation in an emergency. Often, the information has been recorded, it just cannot be found."

In March 1989, Berners-Lee described the blueprint of a platform to enable the people at CERN to share documents easily. The following year, a Belgian systems engineer Robert Cailliau had joined the project, and on 12 November 1990 they published a framework containing the principal concepts behind what we today know as the World Wide Web.

In the below case, Thibodeau (2019) has described an experience management platform produced by Qualtrics.

Product Announcements at Qualtrics X4 Qualtrics produces what it calls an experience management platform, or XM platform. It measures four key areas: employee experience (EX), product experience (PX), customer experience (CX) and brand experience (BX). The firm was recently acquired by SAP for $8 billion.

Qualtrics believes the acquisition by SAP will allow its platform to help close what it sees as a gap between operational data, which includes HR, sales and finance data, and experience management data, whether it's employee,

(continued)

product or brand data. The firm believes that all these activities tie into one another.

"We know happier employees lead to happier customers," Smith said during his talk.

New product announcements packed no surprises. They included more analytics, the use of AI across its product platforms, and new functionality to expand the platform's ability to gather information and analyze it.

This includes a "conversational API" that the firm says will integrate feedback into any chatbot applications. Another new function is called "front-line feedback," which gives employees who work directly with customers a way to provide feedback and submit ideas for improving customer experience.

A new tool on the CX side is "VoiceIQ," which can analyze the human voice in call center interactions. The goal is to understand the sentiment and conduct a "tonality analysis" for insights on customer satisfactions.

New HR-specific tools include "digital listening posts" that allow anonymous employee feedback. It was characterized as an "always on" system that doesn't replace annual or pulse surveys.

Qualtrics also detailed "prescriptive insights" tools that will filter and compare employee data and present managers with potential opportunities for improving engagement.

In the follow-up to the above case, the former President of the USA Barack Obama had shared his experiences on problem analysis and need for a problem-solving culture resulting in transparency and clarity of approaches.

Averbakh, Knauss, and Liskin (2011) have proposed a rights management system for sensitive expertise. The concept accentuates a balanced set of principles for access rights management, where open access and experience reuse have been weighted against requirements for protection of sensitive and confidential information. The principal goal of the approach is to overcome problems in communication and trust for software engineering projects that often are globally distributed. A Wiki-type platform has been selected for implementation.

D'Agostino (2004) presents a case example of Department of Commerce Insider system—and expertise location system, or ELS that directs to an appropriate person, not a stored expertise. According to source, the system is "... freeing up talented people to focus on innovation rather than answering old questions". The paper also states that expertise management is expected to deliver where KM has failed largely due to oversold expectations. The experience of implementing ELS has also shown that sharing of expertise is a counterculture item in many organizations and people generally are reluctant to share what they know because of the fear of redundancy—they might be not needed anymore. However, experts often don't worry about sharing their information because, being the experts, they stay ahead in the field all the time. In addition to this, an ELS-like system saves their time and attention from being bombarded by the same questions again.

All the above examples have several common traits:

- Information that is at the center of the pain points in principal activities;
- Fast and simple delivery;
- Buildup of trust by careful justification of LL:
- Presence of an organizational culture that supports sharing of LL;

6.5 Controversies

Despite the success stories presented above, many sources agree that LL systems have poorly served their goals of reuse and sharing of recorded lessons. As the processes of LL management do not differ much from the processes proposed by knowledge management, the failures are expected to have many features in common as well. In other words, the understanding of factors for success or failure of LL management endeavors may largely borrow from experiences in knowledge management projects. Having this in mind, the following factors for poor success of LL projects are drawn both from the sources on LL and knowledge management. Among the most common reasons indicated for these failures, several directions dominate:

- Improper organizational culture (Frost, 2014; Vuori & Okkonen, 2012; Weber, 2007);
- Inadequate skills of people involved (Frost, 2014);
- Excessive time and effort (Vuori & Okkonen, 2012);
- Motivational issues: information stickiness (Szulanski, 1996; von Hippel, 1994); low motivation to contribute and use (Weber, 2007; Vuori & Okkonen, 2012; Riege, 2006); expert reluctance to codify what they know (Canner & Katzenbach, 2004);
- Poorly organized information—format, level of abstraction (Weber et al., 2001).

Along the same directions, empirical research, presented in more detail at the end of this chapter, has rounded up several LL **controversies**:

- Accumulation of potential trash and deficit of useful experience information—"if I only knew where to look" (*content quality*);
- Reluctance to contribute vs. frustration of unsatisfied needs (*roles*);
- Selection of coverage, format and level of detail for unknown future uses (*format*);
- Constant costs and random benefits, like some kind of insurance against future informing needs.

One of the most important contradictions here, to author's opinion, is a contradiction between the *demand* for LL, which is motivated by need of supporting arguments for decision making and other rational activities, and *supply* that is performed by the people having some useful experience to share. The research on project experience by Petter and Vaishnavi (2008) had shown that only 9% of project members were willing to contribute to the documentation. While the demand is strongly driven by the need to take good decisions and solve current problems, the supply means additional work and is restrained by reluctance. The possible reconciliation of the two could be the creation of a positive feedback loop that is driven by the demand for LL—the very people who have experience this demand could be rather well motivated to join and contribute in the future.

G. Szulanski (1996), discussing the obstacles of knowledge transfer inside organizations, used *stickiness* as a term that defines inertia or reluctance to transfer or receive experience. Szulanski's empirical findings had shown that, instead of motivation, the major barriers to internal knowledge transfer are knowledge-related factors such as the recipient's lack of absorptive capacity, causal ambiguity, and an arduous relationship between the source and the recipient. Discussing the same term of stickiness, von Hippel (1994) had stated: ". . . We define the stickiness of a given unit of information in a given instance as the incremental expenditure required to transfer that unit of information to a specified locus in a form usable by a given information seeker". In addition, von Hippel discussed four patterns regarding locus (location) of innovation-related problem solving:

- When information needed for problem solving is held at some location as sticky information, the problem-solving activity will tend to take place at that location;
- When more than one location of sticky information is required by problem solvers, the location of problem-solving activities will move iteratively among such sites;
- When the costs of such iterations are high, the problem-solving activities might be partitioned into sub-problems, each relating to only one such site;
- When the costs of such iterations are high, efforts may be directed towards investing in "unsticking" or reducing the stickiness of information at some sites.

The last pattern hints at creation of a centralized hub for experience information by finding ways to reduce stickiness and release information held locally.

Shin, Holden, and Schmidt (2001), writing about knowledge management and summarizing the research on knowledge flow, have separated four factors that have the most influence to the knowledge flow: knowledge transferred (content), source, recipient, and context. Although these findings come from a controversial area of KM, we can consider their projection into LL environment, given that in many aspects KM may be perceived as the area, which, at least in the process part, actually deals with advanced information management.

In the same source, a remark is made that ". . . an individual always has a latent incentive to reject knowledge sharing and sell knowledge by leaving an organization. Therefore there is a clear need for rewarding knowledge ownership." In LL context, motivation to share is a cultural issue, and will be discussed in more detail in one of the further chapters of this book.

Another controversy is possible where LLs contradict each other (Gill, 2008). Such situation reduces trust in LL, and should not be left unresolved. If contradicting LLs exist simultaneously, the root of the contradiction may be incomplete context, partial coverage or other situations with incomplete sense. If the contradiction is separated in time, when new LLs reject previous ones, the procedure of unlearning is required. As for participants this might require to abandon a certain part of their common sense, the expected reluctance to do so may be countered by explicit information on changes in context and other conditions.

Cannon and Edmondson (2005) have elaborated on limited resources for managing LL, and noted that having enough time for consulting LLs most likely happens in a hypothetical situation, where an individual has no time limitations. Šimkonis (2016) has noted that the reality of solving problems often brings a choice: to immerse oneself into search for relevant LLs, or to perform an important task for problem solution. The assignment of time for LL analysis **as an alternative** to task execution makes this choice rather complicated. Such decision does not depend upon important features of LL like thoroughness, coverage, relevance and overall usefulness. LL information may be in ideal form and have potential to save time, money and effort, but this is not known at the moment of making a decision. The estimation of the usefulness of LL is possible only by assigning time and attention to evaluate the availability and content of LL.

6.6 Empirical Research on Actual LL Management

The empirical research on the issues of LL or preserved competences is not very abundant. However, in earlier research (Skyrius, 2008) the author, while researching decision support and business intelligence information needs, has surveyed a convenience sample of 250 managers in small and medium businesses on the issues around competence preservation by recorded experience. Some of the questions and collected responses are presented below in Tables 6.4, 6.5, and 6.6.

Table 6.4 Forms of recording the competence information (Skyrius, 2008)

In hard form—formal records and reports in paper or digital media	32
In soft form—as notes, specialist knowledge, experience, opinions	15
Both	202

Table 6.5 The ways of formally recording competence information (Skyrius, 2008)

None	11
Paper documents	166
Computer text (reports)	188
Computer models, procedures, computations	158
Other—e.g., experience in persons' minds	7

Table 6.6 The groups of cases of reusing important decision experience (Skyrius, 2008)

Yes; all the time	81	35%
Yes; sometimes	113	49%
No	33	14%
No opinion	4	2%
Total:	**231**	

Table 6.7 Company size and lessons learned in projects (Chmieliauskas & Šimkonis, 2012)

		Number of employees				
Question	Average	1–10	11–50	51–249	250–1000	>1000
Project lessons are recorded	3.4	**3.2**	3.5	3.5	**2.1**	4.6
Project lessons are discussed informally, without assigning time in agenda	3.9	4.3	3.8	4.0	**3.1**	**3.8**
Project lessons are formally presented (meeting or formal presentation)	3.8	**3.7**	3.8	4.0	**3.7**	**3.5**
It is easy to find earlier project lessons	3.6	3.8	**3.2**	3.8	**2.6**	3.9
At the start of the project, an analysis of former lessons is performed	3.2	3.4	3.5	**3.1**	**2.6**	3.3
Your organization effectively learns from project lessons	3.7	3.7	4.0	3.6	**3.3**	3.7

Table 6.8 Work with lessons learned—approved project management procedures (Chmieliauskas & Šimkonis, 2012)

Question	Approved procedure available	Procedures defined before each project	Projects executed without defined procedures
Project lessons recorded in written	*4.2*	3.1	**2.3**
Project lessons discussed informally, without assigning separate time in the schedule	4.0	*4.4*	3.7
Formal presentation of project lessons (in a meeting or specially arranged presentation)	*4.4*	3.9	**2.8**
It is possible to easily find lessons of previous projects	*4.2*	**2.8**	2.9
An analysis of previous projects is done at the start of the project	*3.4*	3.2	**2.8**
Your organization effectively learns from project lessons	*4.1*	3.5	**3.1**

How the Competence Information Is Expressed, Recorded and Accumulated (Valuable Information About Performed Assignments, Completed Projects, Work Setup, Customers, Markets)?

The responders usually use whatever media and format is most convenient at the moment; organized, hard forms of recorded experience are not numerous.

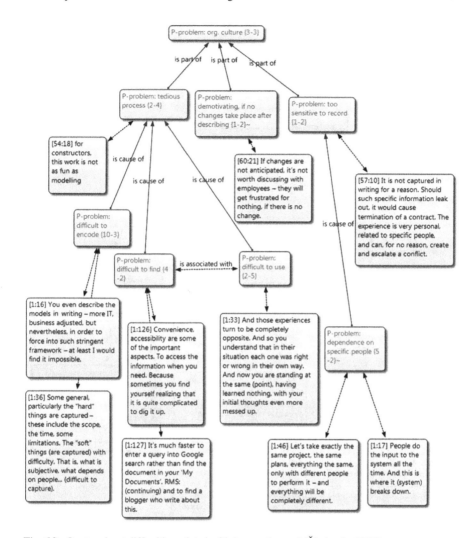

Fig. 6.7 Quotes about difficulties related with lessons learned (Šimkonis, 2016)

Are Important Decisions and Their Development Steps Recorded and Preserved? If So, in What Form?

This information is accumulated as a part of the experience and competence records. With exception of a few cases, the need to produce experience records is undisputed—it provides insurance for future cases, in some cases being the only source of experience information and proof of facts.

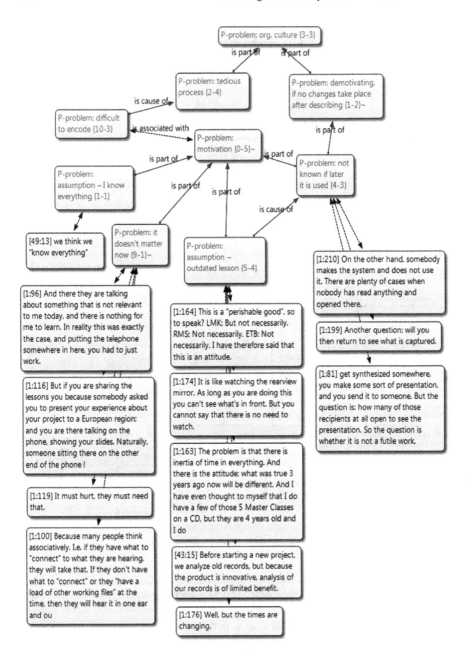

Fig. 6.8 Difficulties related to lessons motivation (Šimkonis, 2016)

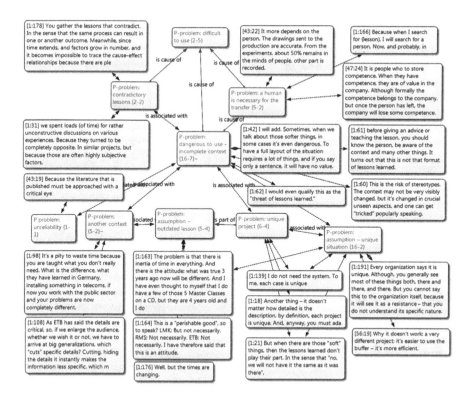

Fig. 6.9 Difficulties in lessons usage (Šimkonis, 2016)

Are There Any Cases in Reusing Important Decision Experience? If So, with What Success?

The response data shows that the reuse of important problem-solving and decision-making experience is of mixed success. The experience records can include anything from free text, anecdotal to hard facts and are recorded in all convenient ways—free text format in digital media, structured format (with some standardized features and values) in digital media, drafts and working papers etc. In practical applications, IT role is seen mostly in arranging, managing structures, imposing standards, allowing easy filtering and retrieval. The reuse of templates, structures, models and other procedural issues is commonplace. Overall, the survey had shown that sensible, demand-driven approaches dominate, and no disruptive, qualitative leap is envisaged.

In 2012, A. Chmieliauskas and S. Šimkonis have surveyed 80 members of Lithuanian project management association to estimate the assumed relationship between the intensity of procedures of LL management and organization size. The respondents have been asked to evaluate the given statements about procedures of

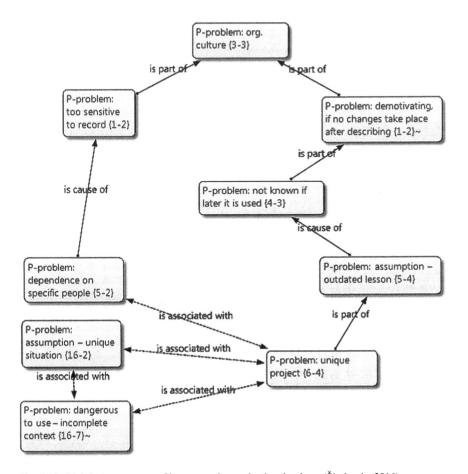

Fig. 6.10 Link between usage of lessons and organizational culture (Šimkonis, 2016)

LL management on the scale of 1–7, where 1 meant "Never", and 7—"Always". Table 6.7 presents the survey results segmented by the number of personnel in an organization (Chmieliauskas & Šimkonis, 2012).

Also in the same survey, A. Chmieliauskas and S. Šimkonis have investigated links between project management maturity and work with lessons learned. The respondents have been asked to evaluate the given statements about nature of their work with lessons learned, reflecting project management maturity via presence of approved procedures, on the scale of 1–7, where 1 meant "Never", and 7—"Always". Table 6.8 presents the survey results (Chmieliauskas & Šimkonis, 2012).

In the table cells, the highest grades are marked italicized, and the lowest are marked bold. It can be seen that in organizations featuring higher maturity of project management the work with LL is more frequent and intense, as compared to organizations with lower levels of project management maturity. This allows for an assumption about relation between work with LL and project management maturity. If, with project management maturity growing, the work with LL would worsen or stay unchanged, we could assume that LL are not important for project management success. For organizations having higher project management maturity, their attention to LL shows that they understand the importance of LL transfer between projects.

In his PhD dissertation, Šimkonis (2016) had researched the use and management of LL in the project management activities. The research involved in-depth interviews with focus groups, collection of spoken responses, their processing using **Atlas.ti** software, and generalization of results. Below are several diagrams showing issues that had emerged during the interviews, and their grouping into topics (Figs. 6.7, 6.8, 6.9, and 6.10).

The indicated difficulties in handling the lessons learned in project management roughly fall into three categories: capturing and recording LL is a tedious process; participants become demotivated if no changes take place after recording LL; some information contained in LL is too sensitive to record and would cause considerable damage if leaked.

Figure 6.8 displays the most common quotes related to the demotivating factors in capturing LL.

The quotes in the Fig. 6.9, reflecting the principal difficulties in LL usage, mostly fall into the following categories: LL are hard to use because of their contradictory nature; they are dangerous to use because of incomplete context; and human presence is required to transfer LL.

The diagram in Fig. 6.10 reflects problems of using LL to certain features of organizational culture: demotivation because of reluctant subsequent use; difficulties in managing and protection of sensitive information; uniqueness and dependence on specific people.

The presented empirical data in this subchapter, although modest in their volumes, allows the following conclusions:

– Discarding expertise is out of question; at least, drafts created around solving a problem are preserved;
– LL activities are recognized as important;
– The potential value of LL, at least from partial reuse, is evident;
– There is a variety of LL forms and approaches; whatever is convenient at the moment will be used; special organized collections are not commonplace;
– Free form records, mostly for "soft" experience, require very few standards and maintain more meaning, but are more difficult and time-consuming to examine;
– Formal, structured records allow for more powerful management and analysis, but squeezing expertise into structured records strips off a part of meaning;
– For larger and more mature organizations, LL are valued more, and structured practices are in place.

6.7 Concluding Remarks

1. Although **the term** "lessons learned" is in use for a substantial time, there is no general agreement regarding the exact meaning of the term. Related terms like "experience management", "expertise management", "knowledge management", "best practices" are often used interchangeably or overlap with the meaning of LL. Some sources, however, point to a clear distinction between LL and just experience, stating that LL are a derivative from experience and have to be actionable. In this work LL are understood as condensed information on experience, formulated with a goal to economically yet reliably convey the meaning of recorded information.

2. The **motivation** of the organization and individual to invest into LL is based on expectations that situation will repeat, and available experience can be effectively reused. However, the creation of LL is not an attractive activity for several reasons. Firstly, it requires additional work, time and effort to contribute. Secondly, it is generally recognized as a boring process that does not seem to create value at the moment of record, as opposed to the activities that it describes in LL. Thirdly, it is complicated by inconvenient record structures and interfaces. This unattractiveness leads to cases when a LL system is implemented and then not being used. The motivation of the use of LL is largely dependent upon organizational and information culture which supports sharing and discussion of LL.

3. **Content**. The content of LL is intended to define key features of a problem and its context. If the content starts being repetitive because of similar cases, triangulation of LL information becomes possible, and rules may be developed as a next step in structuring experience. However, rules as a formatted experience may lose context richness. The changing environment can challenge the existing body of experience and negate previous competencies. Time series of competence changes may be created and used for competence forecasts or other analytical needs.

4. Regarding LL **collection and format**, only part of important experience is recorded as hard data, and the other part stays in people's minds as tacit knowledge. Empirical data from earlier research shows that both structured and unstructured LL records are in use, and the responders usually use whatever media and format is most convenient at the moment. More recent data indicate growing maturity of LL management that introduces standards and structures, and coincides with organization size: larger organizations seem to have a more organized approach towards the management of LL. Structuring of LL information raises new risks: changing environment challenges structures, formats and content; and pushing new information into existing structure may strip important soft meanings from recorded information. Soft information itself is difficult to record, and text format prevails, but important things might stay unrecorded. In addition to this, recorded LL add to information overload if pushed to the users. The records of LL can take different forms—structured records, stories, cases, rules, bodies of

knowledge, varying in their level of structure from collections of drafts to highly structured records with clearly defined standards. An important feature of experience/competence transfer is stickiness—inertia or reluctance to transfer experience.

5. **Issues of using LL**.

Use of LL has shown mixed success—LL have been recognized as useful but limited in their applicability. There's a tradeoff between analyzing LL or taking direct action, especially in the circumstances of time limit. The users of are less likely to listen for somebody else's "broadcasts"—**pushed** experience, but have positive attitude towards their own contributions. There are **accessibility** issues—often LL are hard to locate and retrieve; collections are structured inconveniently. An important element in LL management is indication of the **person**(s) that have been contributing. This data helps in handling sensitive information, or dealing with lacking context—personal communication may add missing elements. Someone's inquiry about a certain experience indicates a real need—and the person possessing this information might be more willing to talk than to record it without knowing whether anybody would read it. Tracking of use and general follow-up is important in estimating the importance of recorded LL.

Regardless of huge advances in information management **technologies**, apparently there are no new tools or techniques specific for handling LL. The existing research leans toward seeing technology not a prime factor for managing experience, but it can be a powerful intermediary in handling meta-information on experience and lessons learned, and be quite effective under the appropriate organizational culture. Many research sources and practical cases agree that a local social network would be a most likely option as a platform for handling LL.

References

Alvesson, M. (2004). *Knowledge work and knowledge-intensive firms*. Oxford: Oxford University Press.

Argyris, C. (1991). Teaching smart people howe to learn. *Harvard Business Review, 4*(2), 1991.

Averbakh, A., Knauss, E., & Liskin, P. (2011). An experience base with rights management for global software engineering. *I-KNOW'11: Proceedings of the 11th International Confwereence on Knowledge Management and Knowledge Technologies. Graz, Austria, September 2011, 1–8.*

Canner, N., & Katzenbach, J. R. (2004). Where "Managing Knowledge" goes wrong and what to do instead. In M. Goldsmith, H. J. Morgan, & A. J. Ogg (Eds.), *Leading organizational learning: Harnessing the power of knowledge*. San Francisco, CA: Wiley.

Cannon, M., & Edmondson, A. (2005). Failing to learn and learning to fail (intelligently): How great organizations put failure to work to innovate and improve. *Long Range Planning, 38*(3), 299–319.

Chmieliauskas, A., & Šimkonis, S. (2012). *A research on project management culture and project manager career situation in Lithuanian enterprises (in Lithuanian)*. Vilnius: Lithuanian Project Management Association.

Cohen, W. M., & Levinthal, D. A. (1990). Absorptive capacity: A new perspective on learning and innovation. *Administrative Science Quarterly, 35*(1), 128–152.

Cyert, R. M., & March, J. G. (1963). *A behavioral theory of the firm*. Englewood Cliffs, NJ: Prentice Hall.

D'Agostino, D. (2004). *Expertise management: Who knows about this? CIO Insight, July 1, 2004*. Retrieved June 20, 2019, from https://www.cioinsight.com/c/a/Trends/Expertise-Management-Who-Knows-About-This.

Doyle, K. (2016). *Xerox's eureka: A 20-year old knowledge management platform that still performs*. Retrieved August 23, 2019, from https://fsd.servicemax.com/2016/01/22/xerox-eureka-20-year-old-knowledge-management-platform-still-performs/.

Easterby-Smith, M., & Lyles, M. A. (2011). In praise of organizational forgetting. *Journal of Management Inquiry, 20*(3), 311–316.

Evans, C. (2003). *Managing for knowledge*. Oxford, UK: Butterworth-Heinemann.

Fleisher, C. S. (2003). Should the field be called competitive intelligence? In C. S. Fleisher & D. L. Blenkhorn (Eds.), *Controversies in competitive intelligence: The enduring issues*. Westport, CT: Praeger Publishers.

Frost, A. (2014). *A synthesis of knowledge management failure factors*. Retrieved August 23, 2019, from www.knowledge-management-tools.net/failure.html.

Gill, T. G. (2008). Structural complexity and effective informing. *Informing Science: The International Journal of an Emerging Transdiscipline, 11*, 253–279.

Hildreth, S. (2005). Rounding-up business rules. *Computer World, 23*, 24–26.

Likierman, A. (2020). The elements of good judgment. *Harvard Business Review*, January–February 2020.

Managing Successful Projects with PRINCE2. (2009). *Office of Government Commerce. Norwich, UK*.

Miller, J. (2000). *Millenium intelligence: Understanding and conducting competitive intelligence in the digital age*. Medford, NJ: CyberAge Books.

O'Donnell, P., Sipsma, S., & Watt, C. (2012). The critical issues facing business intelligence practitioners. *Journal of Decision Systems, 21*(3), 203–216.

Patton, M. Q. (2001). Evaluation, knowledge management, best practices, and high quality lessons learned. *American Journal of Evaluation, 22*(3), 329–336.

Petter, S., & Vaishnavi, V. (2008). Facilitating experience reuse among software project managers. *Information Sciences, 178*(7), 1783–1802.

Riege, A. (2006). Three-dozen knowledge-sharing barriers managers must consider. *Journal of Knowledge Management, 9*(3), 18–35.

Rowe, S., & Sikes, S. (2006). *Lessons learned: Taking it to the next level. PMI Global Congress 2006, Seattle, WA*. Newtown Square, PA: Project Management Institute.

Saran, C. (2019) *The Web turns 30: From proposal to pervasion*. Retrieved August 25, 2019, from https://www.computerweekly.com/news/252459080/The-web-turns-30-from-proposal-to-per vasion?asrc=EM_EDA_109466400&utm_medium=EM&utm_source=EDA&utm_campaign=20190312_The%20web%20turns%2030:%20from%20proposal%20to%20pervasion.

Seaman, C., Mendonca, M., Basili, V., & Kim, Y.-M. (1999). An experience management system for a software consulting organization. *Proceedings of the Twenty–Fourth Annual NASA Goodard Software Engineering Workshop*, December 1999.

Secchi, P., Ciaschi, R., & Spence, D. (1999). A concept for an ESA lessons learned system. In P. Secchi (Ed.), *Proceedings of alerts and lessons learned: An effective way to prevent failures and problems (Tech. Rep. WPP-167)* (pp. 57–61). Noordwijk, The Netherlands: ESTEC.

Shin, M., Holden, T., & Schmidt, R. (2001). From knowledge theory to management practice: Towards integrated approach. *Information Processing and Management, 37*(2), 335–355.

Šimkonis, S. (2016). *Factors influencing the processes of lessons learned in innovative projects. PhD Dissertation*. Vilnius, Lithuania: Vilnius University.

Skyrius, R. (2008). The current state of decision support in Lithuanian business. *Information Research, 13*(2), 345. Retrieved August 28, 2019, from http://InformationR.net/ir/13-2/paper345.html.

Snowden, D. J., & Boone, M. E. (2007). A leader's framework for decision making. *Harvard Business Review, 85*(11), 68–76.

Szulanski, G. (1996). Exploring internal stickiness: Impediments to the transfer of best practice within the firm. *Strategic Management Journal, 17*(Winter Special Issue), 27–43.

Thibodeau, P. (2019). *At Qualtrics X4 Summit, Obama stresses need for facts.* Retrieved August 28, 2019, from https://searchhrsoftware.techtarget.com/news/252459026/At-Qualtrics-X4-Summit-Obama-stresses-need-for-facts?track=NL-1815&ad=926374&src=926374&asrc=EM_NLN_109515547&utm_medium=EM&utm_source=NLN&utm_campaign=20190312_Obama%20speaks%20at%20Qualtrics%20X4%20Summit,%20ERP%20trends%20for%20the%20C-suite,%20SAP%27s%20Qualtrics%20acquisition,%20and%20more.

Tsang, E. W. K. (1997). Organizational learning and the learning organization: A dichotomy between descriptive and prescriptive research. *Human Relations, 50*(1), 73–89.

von Hippel, E. (1994). "Sticky information" and the Locus of Problem Solving: Implications for Innovation. *Management Science, 40*(4), 429–439.

Vuori, V., & Okkonen, J. (2012). Knowledge sharing motivational factors of using an intra-organizational social media platform. *Journal of Knowledge Management, 16*(4), 592–603.

Weber, R. (2007). Knowledge Management in Call Centres. *The Electronic Journal of Knowledge Management, 5*(3), 333–346.

Weber, R., Aha, D. W., & Becerra-Fernandez, I. (2001). Intelligent lessons learned systems. *Expert Systems with Applications, 20*(1), 17–34.

Wenger, E. (2006). *Communities of practice: A brief introduction.* Retrieved August 23, 2019, from https://scholarsbank.uoregon.edu/xmlui/bitstream/handle/1794/11736/A%20brief%20introduction%20to%20CoP.pdf.

Yamauchi, Y., Whalen, J., & Bobrow, D. G. (2003). *Information use of service technicians in difficult cases. Proceedings of CHI 2003, April 5–10, 2003.* FL, USA: Ft. Lauderdale.

Yan, J., Chaudhry, P. E., & Chaudhry, S. S. (2003). A model of a decision support system based on case-based reasoning for third-party logistics evaluation. *Expert Systems, 20*(4), 196–207.

Zahra, S. A., Sapienza, H. J., & Davidsson, P. (2006). Entrepreneurship and dynamic capabilities: A review, model and research agenda. *Journal of Management Studies, 43*(4), 917–955.

Chapter 7
Business Intelligence Technologies

7.1 Overview of BI Technologies

For a considerable time BI has appeared as a technology-driven field with its own flow of technical innovations that enable advanced informing. Virtually all innovations in the field of information management and information systems have been employed for BI activities. Arnott and Pervan (2014) have put together a timeline of decision support and BI systems and technologies (Fig. 7.1). The author of this book has attempted to create a more detailed timeline of the same field (Table 7.1).

The BI technology set features a multitude of advanced tools, and the market for BI technology measures in tens of billions of US dollars by various estimates. There is no common agreement of what constitutes the family of BI technologies, so the market estimates depend upon inclusion or non-inclusion of data analytics, AI and other advanced technologies of relatively recent origin. Over the time frame of BI expansion, technology terms have emerged that are specific to BI—OLAP (On-Line Analytical Processing), ETL (Extract, Transform, Load), SST (Single Source of Truth), in-memory analytics, deep learning and others.

It is important to note that there actually are two sets of BI technologies: the obvious advanced set that every textbook or professional source would list—data warehouses/marts/lakes, OLAP, data mining, modeling, machine learning and the like; and the less advanced and not-so-obvious set that includes many more technologies—less ambitious and visible, but important nevertheless. The most typical representatives of this set are ERP systems, databases with queries, web search tools, Excel models. While these technologies are commonplace and easy to use, their role for BI has been rather under-researched. The full range of analytic tasks includes all of the above, their use depending on the complexity and novelty of the application. It is worth noting that the makers of the large-scale ERP systems have been rather active in the last two decades in offering their own BI extensions to their product family (Microsoft Power BI, SAP Hana), or acquiring the vendors of BI-specific software (IBM—Cognos and SPSS, Oracle—Hyperion, SAP—Business Objects).

© Springer Nature Switzerland AG 2021
R. Skyrius, *Business Intelligence*, Progress in IS,
https://doi.org/10.1007/978-3-030-67032-0_7

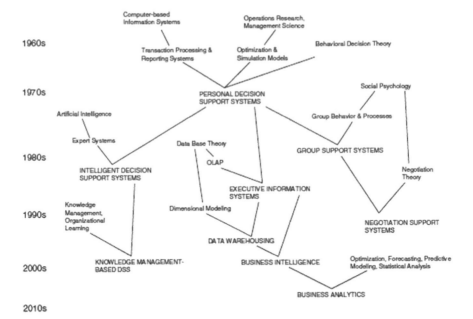

Fig. 7.1 A timeline of decision support and business intelligence systems and technologies (Arnott & Pervan, 2014)

The abundance of BI products on the market can be rather confusing. According to Timo Elliott, an innovation evangelist with SAP (Elliott, 2004), it may not be possible to find a single suite of products that covers all BI needs. It means that different areas of BI activities will be covered by different products developed by different vendors; the products may overlap and require a set of standards for integrating their results. Figure 7.2 below from the same source (Elliott, 2004) shows how several products may overlap and need common sets of standards for the entire BI value chain.

There's no single opinion on the classes of BI functionality, but there is a kind of rough consensus about the general BI technology categories by their function:

1. Data collection and storage—ETL, data warehouses, data marts, data lakes etc.
2. Self-service tools; query and reporting tools
3. Multidimensional analysis, OLAP
4. Data mining for large data sets
5. Big Data analytics; AI and machine learning
6. Text analysis and mining
7. Modeling and simulation
8. Presentation and visualization
9. Communication and collaboration platforms

Table 7.1 Composite timeline of decision support and business intelligence technologies and innovations

Year	Research	Technology	Applications
1865	Richard Devins uses the term BI in his book "Cyclopaedia of Commercial and Business Anecdotes"		
1945		First computer, the ENIAC	
1950	Alan Turing proposes a test for machine intelligence		
1947	George Danzig publishes the simplex method		
1952			George Danzig—first linear programming applications
1955	Jay Forrester proposes quantitative models for DS; the term "artificial intelligence" is first coined by computer scientist John McCarthy		
1956		First hard disk drive for mass data storage	
1958	Hans-Peter Luhn mentions BI. Frank Rosenblatt creates Perceptron, a first neural network algorithm at Cornell Aeronautical Laboratory		
1959	Machine learning term is launched by Arthur Samuel, then researcher at IBM		
1963		First DBMS, Charles Bachman's IDS with General Electric (some sources indicate 1964)	
1964	A chatbot named Eliza is created at the MIT AI lab	IBM System/360—a family of compatible computers is introduced	
1965			Doug Engelbart's first prototype of a GDSS. Development of a first ES, Dendral, in Stanford. Development of BMDP (Bio-Medical Data Package) in UCLA.
1968			SPSS, a statistical analysis software package, is launched

(continued)

Table 7.1 (continued)

Year	Research	Technology	Applications
1969		Foundations of Internet laid in ARPANET project	Cognos, a BI software company, is founded
1970	First developments of would-be DSS. Michael Scott Morton's PhD on decision support		
1974		Xerox develops Ethernet local area network protocol	
1975	A term "Decision support system" is launched		Launch of Express by Information Resources, a first OLAP product
1976			Jim Goodnight founds SAS ("Statistical Analysis System"), until then development from 1966 at NCSU
1977	Steven Alter defines first typology of DSS	Apple II personal computer released	
1979			Launch of the first spreadsheet, VisiCalc
1981		IBM PC personal computer released. Xerox Star workstation released containing the first commercial graphical user interface.	Hyperion, a BI software company, is founded
1982	Sprague & Carlsson book establishes DSS as a class of information systems		
1985			Microsoft Excel 1.0 launched
1989	Howard Dressner defines BI—"Concepts and methods to improve business decision making by using fact-based support systems"	SQL data querying language is launched	University of Arizona launches Ventana Corporation to market Plexsys technology for group decision making. CMU uses neural networks to steer autonomous cars.
1990			Business Objects, a BI software company, is founded in France
1992	Bill Inmon publishes his book "Building a Data Warehouse"		Crystal Reports, a reporting software, is released
1993	Term OLAP is coined by E. F. Codd		QlikTech launches in Sweden
1994		Commercialization of the Internet	Netscape Navigator browser is released

(continued)

Table 7.1 (continued)

Year	Research	Technology	Applications
1995	Montreal hosts first international conference on data mining and knowledge discovery in databases		
1996	Ralph Kimball publishes "The Data Warehouse Toolkit"		
1998	Big Data term is launched by John R. Mashey, chief scientist at Silicon Graphics Inc.		
2000	Daniel Power defines second typology of DSS		
2001	Term "SaaS" first appears in an article in Software and Information Industry's eBusiness Division		
2002			Wayne Eckerson founds TDWI (The Data Warehouse Institute) Research
2004			Facebook launched
2006		The rise of the cloud and mobile BI	Apache releases Hadoop 0.1.0
2007		Apple's iPhone is launched	
2010		SAP launches HANA	
2012			Barack Obama uses predictive analytics for election campaign
2015		Microsoft, Amazon, Hewlett-Packard and Accenture launch machine learning and advanced analytics platforms	
2016			Insight development goes beyond professional statisticians
2017			AlphaGo software beats the world champion at Go

(Source: author)

The rest of this chapter is going to be organized along the above groups, with addition of deployment issues in the end. There are other categories that are often mentioned lately: *augmented analytics* that use AI methods to support analysts and other users with procedure automation; *predictive analytics* that use historical data to produce future projections, and, as such, do not differ much from long-known forecasting techniques; *prescriptive analytics* that suggest decision alternatives and

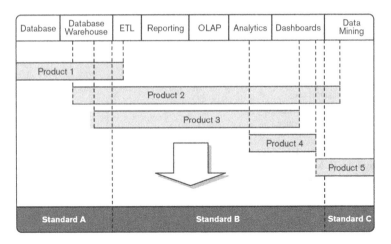

Database	Database Warehouse	ETL	Reporting	OLAP	Analytics	Dashboards	Data Mining

Product 1

Product 2

Product 3

Product 4

Product 5

Standard A   **Standard B**   **Standard C**

Fig. 7.2 Overlapping of BI functions, products and standards (Elliott, 2004)

action courses; and others. There is no doubt that at the time of writing more new categories are arriving and will be the subject of discussion and evaluation in the coming years.

7.2 Data Collection and Storage

The subject of this subchapter is technologies for data collection and storage with BI functions in mind—databases, data warehouses, data marts and data lakes, data vaults, other collections. Non-traditional collection technologies—NoSQL, column databases and other less common forms of data collection also belong here as mutations of traditional data models with enhanced analytical capabilities in mind. The importance of capture of exact records ("no fact stays unrecorded") and their quality is undisputed. The recent growth of automatic data capture by variety of capturing devices operating in the environment dubbed Internet of Things creates both new opportunities for additional insights and concerns in handling those opportunities.

Over the years, data collections have always been useful for analytical purposes. In the pre-IT times, technologies like document collections, file cabinets, Rolodexes were just the few forms of organized data for whatever need to use it might have occurred. The ubiquitous databases of today with volumes and handling capacity undreamt of in the pre-computer era have provided the first stages of analytical capability with flexible queries, appends, joins and other content-enhancing functions. The foundations for the current wave of interest in data analytics largely lie in these everyday database technologies, and the American Airlines SABRE system makes a good case for this point. While the original system has been intended to handle the growing demand for air travel and ticket sales, eventually opportunities to gain insights from the data collected in the system came up and significantly boosted the competitive potential of the system owner—American Airlines.

In 1992, one of the creators of data warehouse concept, William Inmon, has presented one of its first definitions (Inmon, 1992): A *data warehouse* is a subject-oriented, integrated, time-variant and non-volatile collection of *data* in support of management's decision making process. In this early definition, the accentuated orientation of DW towards decision making, or satisfaction of the complex part of management information needs spectrum, is obvious—there are important pieces of valuable information to discover in the accumulated data, and it requires, among other things, a more analytically capable data organization technology than DBMS. When compared to DBMS, data warehouse technology is intended to process business history instead of a business snapshot in DBMS, and is much more enabling for discovery (March & Hevner, 2007).

A key procedure associated with development of a data warehouse is the ETL procedure. According to Verhoef, Kooge, & Walk (2016: 93), *extraction* covers data source validation (are these the correct and required data?), frequency—how often and how fast; typically real-time is not technically and financially feasible. If real-time insights are needed, business rules and algorithms should be used. *Transformation* includes coding, recoding and error handling where possible. *Load* phase might use staging to prepare data to be uploaded ("lift to a loading platform"); this data stays if the uploading goes wrong. Loading is usually done by appending and less by overwriting.

The two dominating DW models are Inmon and Kimball models, named respectively after William Inmon and Ralph Kimball—the two authorities in the area of DW that have laid most of the DW foundations for today.

In Inmon approach, DW development starts in a top-down mode with a corporate data model reflecting key entities in business—products, customers, suppliers, employees etc. Normalization is executed throughout the model, starting with lone entities and spreading to physical model, with a goal to avoid data redundancy as much as possible. The single truth version, or SST (Single Source of Truth) is managed at this level. The main advantages of Inmon approach are (Inmon, 1992):

- DW serves as an SST and repository of standards due to its scope encompassing all the data in an organization;
- Standardization helps solving many data quality issues;
- Centralized nature prevents data and schema update anomalies are because of the rigid centralized standards;
- Approach enforces reporting transparency across the enterprise.

The disadvantages of Inmon approach are similar to those of any centralized systems:

- It is implemented by large projects whose development time is substantial;
- Data model may become overly complex as the number of data tables grows over time;

- Project implementation and management requires a significant team of data modeling and data management specialists.

The Kimball model starts in a bottom-up mode with key business processes, and the data from these processes and related operational systems are loaded into a dimensional model. This model is built around star schemas and is not normalized—this is the key difference between Inmon and Kimball models. The advantages of Kimball model are:

- Quick development and time-to-operation;
- Environment is less complex and requires less staff to operate;
- Performance is effective, providing for easy creation and use of data marts and drill-across operations using multiple star schemas.

Disadvantages, again, are similar to those of any decentralized system:

- Complicated to conform to SST;
- Problems in updating data and structures; integration of legacy data can be complicated.

Altogether, the principal differences rather coincide with the general differences between centralized and decentralized systems. These differences impact the initial development time of the DW, and the future flexibility to update DW design. Hybrid models exist that attempt to combine the advantages of both models. Rangarajan (2016) relates the choice between the two models to organizational culture of either long-term returns or quick value.

Data marts. A data mart is a local-scale data warehouse, or a low-level repository that contains domain-specific information in the activity area of a single business unit. Hence, the typical smaller-scale project advantages apply: a data mart may be developed quickly because of its limited scope; it is easier to manage and more agile; and the cost of managing a DM is low. The small scope and simplicity also introduce clarity in understanding business requirements, so the chances for a successful showcase DM project are significant. DM relate rather harmoniously to R. Kimball's model, which sees DW as a cluster of related DMs.

Data marts are not without their own risks. A considerable risk emerges when growing data marts: the entire whole of data assets thus may migrate into silos that will aggravate data integration problems in the future. A single source of truth, the significant part of DW ideology, may become meaningless if overall DW consistency is improperly managed. A smoothly operating data mart for one business unit, built without inter-unit data integration consistencies in mind, prevents the analysts from seeing the organization-wide view or analyzing the dynamics of several related business processes side by side.

Data lakes is a concept opposite to data warehouses: whereas data warehouse data are carefully prepared, cleansed and loaded, data lakes store mostly raw or mixed data, structuring only what's necessary. Where DW is a purified, laboratory-grade set of records that is easy to manage and manipulate but has lost parts of provenance, coverage and contextual semantics in the process, DL is a messy real-life reflection that is harder to manage, but maintains much more context. Under DL, raw data is loaded into a lake and transformed if a need may be, which is an alternative to the ETL process. Fine-grained micro-services for specific business functions can be easily added or replaced. DL also can serve as a staging area for DW.

While DL approach has its pros and cons, data lakes can be too messy for reaching structured data. Easy access and consolidation, as well as the overall flexibility are doubtful. One of the most often mentioned risks with DL is DL becoming "data swamp", "data dump" or "data graveyard" with chaos and no structures to manage.

Data vaults. Data vault is one more form of data repository, built on assumption that all data in an organization is relevant data and should be kept together. According to Dan Linstedt (2002), the creator of the data vault form, it is a hybrid approach encompassing the best of breed between third normal form (3NF) and star schema. The data vault has to accept all data, and data is interpreted when it is extracted from data vault or reports are created.

The data vault approach is based on three types of entities: hubs, links and satellites. Hubs contain a unique table of primary keys that identify objects participating in everyday operations, e.g., customers, products, employees, suppliers, components. Links join hubs with other hubs by containing the keys of all hubs linked via this link. Satellites contain descriptions of objects contained in the hubs—names, prices, other features. A hub is always connected to at least one satellite. All three types of objects have two mandatory fields—time/date stamp indicating the moment of the origin of this record; and origin field indicating the source of record.

The principal benefits of the data vault approach are, based on (Kranc, 2017): better performance; increased usability; better traceability and auditability; and support for isolate, flexible and incremental development.

Key performance indicators. For advanced business monitoring, DW approach is expected to enable effective composition and tracking of essential key performance indicators (KPI). Effective KPIs reflect the rate of the key business processes and functions that are the most important in meeting strategic goals and performance targets. Actually, DW design should be driven by the needs for certain KPI (Jeusfeld & Thoun, 2016) and the subsequent set of requirements for DW organization. This is not so easy to achieve because the structure of operational data sources limits the possibilities of a DW.

Most sources agree that KPI are highly aggregated indicators that project their rates against against top-level goals. Each important business aspect should have no more than 1–2 KPIs. At the same time, KPI should have a hierarchy by the usual strategic, tactical and operational levels. Such hierarchy is able to allow "drill-down" diagnostics and traceability of important issues.

While the available experience of DW use has been mostly positive, there have been criticisms aiming at DW projects being time-consuming, and DW themselves as limiting and inflexible. While such criticisms are expected for all kinds of large-scale systems and projects, there are some DW-specific pain points. In practice, the support for KPI hierarchy and fine-tuning of aggregation levels across the organization adds issues to the already complicated data integration issues when developing a DW. Different business units may have their own KPIs that might or might not be harmonized with other units. Furthermore, the issues of KPI reconciliation may be aggravated by inevitable changes in KPI mix, testing the flexibility of data models.

7.3 Multidimensional Analysis, On-line Analytical Processing

On-Line Analytical Processing (OLAP) technology is based on relatively simple approaches that serve to answer many relevant business questions. OLAP genesis stretches from DB queries, through SQL to OLAP-specific environments and tools. The roots of OLAP technology lie in report generators (RPG) and database queries. Some sources (Berry & Linoff, 2004) state that OLAP technology is a part of data warehousing technology. Although it is true that OLAP becomes a powerful tool if it uses a data warehouse designed with key business dimensions in mind, the source data need not necessarily be loaded into a data warehouse; a relational database may also run OLAP queries. However, in this case response efficiency may drop significantly.

OLAP is also often mentioned when discussing data mining methods. While data mining toolset is far more capable than just multidimensional data analysis, OLAP strength lies in the space of clear business dimensions that is understandable to the business user and allows easy projection of business questions onto data. For example, cluster detection is a classical data mining task, but the detected clusters might be hard to explain in a business sense. OLAP dimensions can help by defining specific features of the clusters and making sense of their origins.

With the rise of analytical information needs, it soon became apparent that the "workhorse" of data management infrastructure—the database management system (DBMS) cannot be equally effective for everyday transaction handling and servicing analytical queries at the same time. Analytical queries launched upon an already overburdened database congest everyone else's work, including vital transactions, and the query itself will be completed in minutes or hours. As a result, a clash between transaction processing and analytical processing came up: while DBMS services proved adequate for OLTP (On-Line Transaction Processing) systems servicing everyday operations, the emerging OLAP activities, being rather resource-hungry, strangled OLTP processes. Besides, rigid DBMS structures, normalized and optimized for transaction loads, limit analytical potential of data use.

OLAP technology has been developed with satisfaction of analytical information needs as the primary goal. One of the strengths of OLAP technology is the focus on business dimensions—time, place, product etc., that are clear to the business user. Important business measures and ratios can thus be quickly implemented and refined. The dimensions define a data structure that is a multi-dimensional cube made up from elementary cells; each cell contains a fact and its dimensions, and is located at a point defined by the instances of its dimensions. For example, a shipping order may have a following structure:

Order date
Delivery date
Client
Location
Product
Quantity
Unit price
Discount size

In this example, the facts are: quantity, unit price and discount size; and the dimensions are: order date (time dimension), delivery date (time dimension), client (client dimension), location (location dimension) and product (product dimension). Some dimensions contain hierarchy levels, as they can be subdivided or aggregated into different units—e.g., time hierarchy handles days, weeks, months, quarters and years; location hierarchy—street address, municipality, metropolitan area, region, country; product hierarchy—product group and subgroups by some agreed coding system like EU product classification system; customer hierarchy may include customer segments and sub-segments.

There are several types of generic OLAP operations that users can perform (Biscobing, 2019):

- **Roll-up,** also called *consolidation* or *aggregation,* summarizes the values of data and produces totals along the dimension.
- **Drill-down** allows analysts to navigate deeper into the dimensions of data, like drilling down from "month" to "week" or from "region" to "municipality" to examine sales growth for a product. This operation helps to focus on a root of a problem.
- **Slicing** enables to take a subset of data along single instance of a certain dimension, such as "sales in November 2019."
- **Dicing enables to take a data subset from several dimensions; for example,** "sales of composite fishing rods in store X in the last quarter of 2018."
- **Pivot.** New views of data can be produced by rotating the data axes of the cube.

Although OLAP technology is not considered advanced, it has proved helpful in many cases supporting answers for business questions that are of average complexity, or discovering issues that require more attention and deeper analysis.

7.4 Use of Simple Self-Service Tools: Role of Excel

Arena, Rhody, & Stavrianos (2005), discussing advanced analytics in a joint context of decision support and business intelligence, stated a growing need for self-service analytical applications, and named specific challenges related to these applications:

- **Diverse problem set and specialized functionality**—the broad variety of business processes and problems limit the number of users for a specific tool; universal applications with customization are of limited applicability;
- **Highly technical algorithms**—they may be too complicated for non-technical users; even if this complexity is shielded from users via friendly interfaces, users still need traceability to have confidence in the results;
- **Advanced information delivery**—the analytic applications must possess intelligent, interpretive delivery of results; output should be designed with users' attention management;
- **Evolving requirements**—applications should be agile enough to adapt to: changes in data sources; changes in analytic methods; and changes in business processes.

The role of the self-service BI technologies is to serve the analytic side of everyday technologies—MIS, ERP and others. Such everyday analytical functions, performed mostly by the front-line employees, are identified by the terms of *embedded analytics* or *operational intelligence* (OI). Its applications are mostly used by operations level employees, e.g., salespeople, contact center workers, supply or distribution operatives, and impact their efficiency by enabling more confident and faster action, better decisions, and overall gaining more independence. This expected improvement comes from timely and properly engineered intelligence information. In addition, operational intelligence uses automatic alerts to specified events or conditions where applicable. Lately, such automation of rules and responses is receiving more and more attention as the area where reasonable investment into automation and AI techniques promises significant payoffs.

Operational intelligence has evolved from traditional BI querying and reporting with simple techniques, fast learning and smarter work environment where current information needs are met faster. As technology developments continue, tools seen as rather sophisticated yesterday are now accessible to operational users to perform their own analysis. Environments often are implemented as real-time or near-real-time BI systems that can combine current monitoring of data streams, data integration from different sets, data analytics. For efficient operational informing, environments use visualization and dashboards to reflect operational metrics and KPIs. Threshold values and alerts are typical, managing user's attention, and minimal context is required to make sense of the situation around alert. A tiered structure of such environment, as described in Chap. 4, is possible where the first tier would contain key data and tools for instant clarification, with the more powerful functions in the background.

Operational intelligence creates most value in fast activities where rapid reaction is required. Customer analytics is one such area where data from various sources like clickstream data, actual transactions, customer questions or requests are combined and analyzed to customize marketing campaigns. At the same time, this stream of data is a rather reliable source for detecting emerging trends, creating additional value on a higher step of analytical activities. Another typical area is stock market trading where huge volumes of data have to be analyzed in real time, and exactly this area has been a subject of huge efforts of automation over the last couple of decades. Mobile communication companies track data from their channels and communication hubs in real time to analyze the usage patterns for better-tailored marketing campaigns and customer proposals.

A special type of OI applications has emerged due to the growth of the *internet of things* and its array of assorted sensors capturing data from a myriad of spots of interest: processing and manufacturing machinery, energy and communication networks—virtually any equipment that allows the digitization of data on its behavior. For complex technical systems, the real-time availability of this data supports predictive maintenance to deal with likely problems in a proactive way. The very activities of IT in any business also generate volumes of machine data in server, network and website logs to be constantly monitored for issues like congestion, failures and security threats.

Apart from rushed real-time OI systems that serve users under time pressure, there are other, more relaxed environments. Their needs for operational insights, however, remain the same minus urgency.

A typical representative of such environment is call-center applications that bring the vital client information at the operator's fingertips as soon as the caller is identified.

7.4.1   Queries and Reporting

Queries and reporting traditionally are two complementary analytical functions to assist users outside the analysts circle. Because of their relative simplicity, queries and reports do not need extensive training of the users, and blend nicely into the fabric of everyday activities if the interfaces are simple enough for an average non-technical user. A query searches a data collection against some conditions and returns a set of data in a raw format, while a report presents data in a readable, easy-to-use format. Contemporary querying tools may have formatting functions for better usability of results, and reporting tools may have querying functions that allow specifying conditions for data selection. In any case, for self-service BI the use of simple query and report tools is commonplace. This simple stage of initial analysis is often enough to disclose underlying reasons for business results.

Although most ad hoc reports and analyses are meant to be run only once, in practice, they often end up being reused and run on a regular basis. This can lead to unnecessary reporting processes that affect high-volume reporting periods. Reports should be reviewed periodically for efficiencies to determine whether they continue to serve a useful business purpose. Self-designed reports also may become a source of inconsistencies; and, once unnoticed and reused, inconsistencies may multiply.

7.4.2   Spreadsheets

Although business intelligence has been a testbed for advanced (and rather expensive) analytical technologies, a simple and cheap alternative has existed since the emergence of spreadsheets. The pervasive nature of spreadsheet use is easy to explain—unlike any traditional information system, spreadsheets arrived on the back of personal computers as a significant part of IT democratization wave, and were adopted voluntarily rather than "implemented" by decree. Features like simplicity and low starting threshold made spreadsheets a welcome personal productivity tool for anyone working with structured data, and their spread is seen by some sources as a way to pervasive BI (Tront & Hoffman, 2011). The early brands VisiCalc, Lotus 1-2-3 and Borland Quattro eventually gave way to Microsoft Excel, the latter currently being synonymous with any mention of spreadsheets. The widespread adoption of spreadsheets created significant advantages as well as accompanying challenges, and the advantages can morph into challenges in quite a few occasions.

The *advantages* of spreadsheets are more or less clear:

- A democratic tool that provides considerable freedom to its users independently of rigid common systems;
- Simple and easy to learn—there's no need for IT education;
- Ubiquitous software—a "Swiss army knife" that keeps getting more powerful, with features approaching those of specialized BI tools;
- Ad hoc analysis may be performed by users anytime as needs arise, without the need to wait for static, regular reporting of centralized systems;
- Such fast analysis may enable users to get the insights as fresh as possible;
- Common platform and standards to facilitate sharing;
- Reduced load for centralized systems.

However, the same advantages may turn into *risks*. The solo nature of spreadsheet use means that the content is extra-systemic, or outside other information systems in an organization, and therefore rather tricky to manage. Spreadsheets are not designed for collaboration, so joint work on a single spreadsheet model is complicated. Ease to learn, combined with lack of experience, may lead to errors in formulas and formatting, or to reckless use of advanced features without really understanding the dangers. The errors are hard to manage and validate. The ubiquity may lead to beliefs that every user understands spreadsheet techniques in the same way, resulting in severe inconsistencies and mismatched results. The growing sophistication of spreadsheet features grows the risk of inconsistency accordingly. Same can be said about common platforms—while spreadsheet standards cover the basics, the individual designs by separate users, if ungoverned by rules or conventions, tend to drift away from each other, thus aggravating the future integration. The extra-systemic nature of spreadsheets raises serious security issues as well.

The above risks have caused significant concern and talks of "Excelization" as a troubling phenomenon that has to be put under control. However, the recent trends coming up in discussions at BI conferences or in virtual space tend to lean towards more harmonious co-existence of spreadsheets and other IT functions.

7.5 Business Analytics

The very term "analytics" covers a vast array of approaches, methods and tools to process chaotic sets of data and discover useful information. Regarding business data, a scale of analytical methods can be drawn by their complexity and sophistication from the simplest to the most complex. Simple analytical procedures—grouping, totals, drill down, roll up, slice and dice, pivot—are the domain of queries, reports and OLAP technology. Contemporary analytics pick up where queries and OLAP stop, although many sources name OLAP as one of the methods employed by analytical activities. Advanced analytics are at the complex end of complexity dimension, and, according to Gartner (Advanced Analytics, n.d.) use extensive statistics, data/text mining, machine learning, pattern matching, forecasting, visualization, semantic analysis, sentiment analysis, network and cluster analysis, multivariate statistics, graph analysis, simulation, complex event processing, neural networks. In BI value chain or insight building activities, analytics may be seen as a set of manufacturing tools that produce sophisticated results from elementary data with significant aggregate meanings.

Some sources reflect complexity on a past-future dimension: descriptive analytics on historical data are seen as simple, and predictive analytics, together with prescriptive, are complex. However, analysis of historical data may be rather complex too; and some prediction methods are of medium sophistication.

At the time of writing this book, business analytics are strongly on the rise, and currently are challenged by its popularity—flashy terms abound that distract attention and dilute the meaning of purpose and value. There are numerous study programs for business analytics, data science and the like; job market is bristling with lucrative positions in the area; research and professional sources signal the emergence of new journals and conferences, and industry experts forecast the further growth. There is a multitude of sources that discuss analytical methods in great detail, so in this book the details of analytical methods will be given an overview, as technology details are not the primary focus. There might be certain omissions, but, as for any domain, no list of techniques and approaches is ever final.

7.5.1 Data Mining

The concept of data mining has emerged in the mid-nineties of the XX century, joining together statistical techniques, database queries, artificial intelligence and visualization methods. Although the elements were not exactly new, the concept gained momentum as the new form of analytical technology, intended to discover new, previously unknown and useful information in large data collections. Several prerequisites that weren't available until then have attracted the attention of the user community (Sharda, Delen, & Turban, 2014):

- Business need for tools to compete that are based on better understanding of business environment, customer behavior in the first place;
- Large collections of data available for analysis;
- Storage and processing technologies getting ever more powerful and cheaper;
- Integration of data around objects of interest, such as customers, suppliers, markets, processes etc.

A universal process defining data mining phases, CRISP-DM (Cross-Industry Standard Process for Data mining), has been developed in the last decade of XX century. The value of data mining mostly comes from its ability to reach out further from the point where traditional data querying methods stop, and discover hidden information using sophisticated intelligent methods like neural networks or genetic algorithms. Some sources (Adriaans & Zantinge, 1996) state that around 80% of interesting information contained in a database can be extracted using traditional queries, and the remaining 20% require sophisticated and powerful approaches like data mining.

Data mining technology is one of the early users of machine learning techniques that extract rules from observations and adjust their accuracy with newly received data. Important anomalies and patterns are expected to be discovered using both supervised and unsupervised learning. In supervised learning, classes or conditions that represent a hypothesis or a business question are defined by the user; while in unsupervised learning, classes, clusters or patterns have to be discovered by the technology itself.

The methods of data mining fall into following groups:

- *Classification*—use of user-defined classes that are derived from data attributes belonging to objects of interest to assign the newly-arrived objects to a certain class; an example is an assignment of a new client to a client category according to behavior. The more prominent algorithms are J. R. Quinlan's **ID3**, **C4.5** (Wu, Kumar, Ross Quinlan, et al., 2008) and **CART**—Classification And Regression Tree (Breiman, Friedman, Olshen, & Stone, 1984).
- *Clustering*—segmentation of objects into partitions due to their similarity or proximity; example—estimation of dominating customer segments. The more prominent algorithms are **K-means** for non-hierarchical clustering, and **ROCK** (Guha, Rajeev Rastogi, & Shim, 2000) and **BIRCH** (Zhang, Ramakrishnan, & Linvy, 1996) for hierarchical clustering.

- *Association*—discovery of affinities or patterns based on repeating groups of items from a common collection, or association rule mining; a typical example is "basket analysis" in retail. Well-known algorithms are **Apriori** (Agrawal & Srikant, 1994) and **CARMA**—Continuous Association Rule Mining Algorithm (Hidber, 1998).
- *Forecasting*—estimation of future data values using regression techniques; example—prediction of demand for goods to optimize stock levels.

Although not a separate group of methods, the below two groups, to author's opinion, have to mentioned here because of their importance:

- *Anomaly detection*—identification of unconventional events or observations that differ significantly from the rest of the data; example—detection of unconventional behavior. Anomaly detection may use all of the above methods.
- *Visualization*—use of graphical techniques to support interpretation and evaluation of results; they can be used with virtually any analytical task or method. The use of graphics is limited by the two-dimensional space on monitor screen, but the long-awaited advent of 3D monitors should raise visualization power to another level.

An important class of data mining tools are *artificial neural networks* (ANN)—one of the successful developments in the field of artificial intelligence over the last several decades. They are used across many data mining methods like classification, clustering, prediction and association. Although being relatively simple in their concept and technology, ANN have two profoundly important features (Zhang, 2008): parallel processing of information; and generalizing and learning from experience.

Genetic algorithms are another important class of advanced data mining tools. Although various sources claim that genetic algorithms are an alternative to traditional machine learning methods because of their ability to solve problems that were deemed impossible to solve because of the enormous search space, their importance seems to be diminished in a current data analytics landscape (Skiena, 2010). Several data mining application examples are presented in Table 7.2.

In its ways to discovery, data mining in business borders with scientific research—both strive to notice weak yet important signals, and in both cases there is substantial concern about how believable or reliable the results are. Detected correlations do not automatically imply causality, so additional steps of sense-making will be required to discover or justify causality.

7.5.2 Big Data Analytics

Although the notion of Big Data is somewhat muddled, and has many things in common with databases, data mining and data analytics, several specific features may be helpful to have a clearer definition of what is understood by the term. The

Table 7.2 Examples of data mining applications

Application	Method and gains
Basket analysis in retail	Analysis of combinations of goods purchased together to develop a set of rules for most common combinations and leverage this information for promotion or other purposes.
Churn prediction for mobile operator	Analysis of a set of former customers that have switched to competitor to determine specific features of their behavior. The results allow development of a diagnostic profile of a candidate for leaving, evaluation of current such candidates, and possible actions.
Stock trading	Guidance of investment decisions by analyzing market data together with possible join with additional other data collections, information from public and social media.
Recognition of stolen credit cards	Detection of unconventional behavior in the use of a card that does not conform to usual patterns of use, and fast action—contacting the user or blocking the card.
Loan application analysis	Evaluation of bad debt risk of current applicants, based on historical application data and classification of current applicants.

Source: author

ability to deal efficiently with vast collections of digital data has been around since the early days of computing. The advances in IT and growing capacity of data storage gradually ensured that larger and larger collections of data could be processed relatively easily and with low costs. However, the growth of data collections past some point makes it difficult to handle with traditional database technologies. As well, the variety of sources and formats plus growing role of unstructured data pushes the data handling rules outside Codd's model.

The common definitions of Big Data usually include vast amounts of data, or *volume*; heterogeneous and messy data in a multitude of formats, or *variety*; and many transactions (hundreds to thousands or even millions for a global business) over a short period of time, or *velocity*. Some sources go even further and add two more Vs: *veracity*, or reliability of data; and *value*, which speaks for itself and probably should be placed first. Accordingly, technology issues emerge that address key differences between traditional data and Big Data (although no one can tell where exactly traditional data ends and Big Data begins!):

- Volume—for Big Data, technologies have to facilitate massive parallel processing across multiple storage platforms;
- Velocity—for Big Data, technologies have to process data in near-real time;
- Variety—for Big Data, technologies have to accommodate heterogeneous data in a variety of models and platforms.

Big Data analytics uses discovery principles taken from data mining, AI techniques to expand capabilities, and massively parallel processing to handle the volumes. Specific software tools (e.g., Apache Hadoop) have been developed to process large data collections in clusters of computers that may contain hundreds or thousands of computers. The key components of Hadoop technology are Hadoop Distributed File System (HDFS) and MapReduce module, who handle the placement of data, distribution of data analysis task (Map), and assembly of processed information (Reduce).

The use of AI approaches has led to high expectations in data analytics; however, human presence in the analytical process is considered vital by numerous sources (e.g., Berry & Linoff, 2004). One of the important reasons for human presence is the iterative nature of analysis: the system has to provide feedback to the users and accept fast modifications of inquiry parameters.

As any area with huge expectations brings up its own set of controversies, Big Data analytics is no exception. Strange results like spurious correlations or false positives are specific to Big Data. In huge amounts of data, statistically significant correlations will be found that are impossible to explain, introducing confusion between correlation and causality. This is one more reason to have human presence in the analytical process. The spread of Big Data approaches has also sparked debate on ethical issues like privacy and data ownership. Having quickly mastered Big Data analytics, businesses leveraged access to varied and scattered data sources. And, not for the first time in history, business moved faster than legal bodies managed to draw regulation rules.

7.5.3 Descriptive, Predictive and Prescriptive Analytics

The emergence of predictive and prescriptive analytics suggests a chain: descriptive—predictive—prescriptive analytics. The latter two stages contain models that use heuristics and rules derived by analysis of past data (descriptive analytics), perform forecasting and develop suggestions to act. Sometimes two more types are mentioned—diagnostic analytics and planning analytics, although they are very close to descriptive analytics and predictive analytics respectively.

It is worth mentioning that the field of decision support has for long time intensely used all three (or more) types of analytics to deepen understanding of the problem under decision. The recent revived interest in decision support can probably be explained by the rise of analytics and a very close overlay of the two fields.

Descriptive Analytics

Descriptive analytics answer the question "What has happened?" and are based on statistical methods that analyze past data to produce insights of what exactly happened, what has been going on, or what is a current state of affairs. The majority of analytics performed in business still belongs to this area. Reporting can be seen as a simple form of descriptive analytics. The results are reliable because they have been based on facts, given that correct analysis methods have been applied and the quality of raw data is acceptable. In descriptive analytics, KPI and other important indexes have a key role, bridging a gap between basic reporting and deep diagnostics.

Predictive Analytics

Predictive analytics is an area of data analytics that deals with extracting information from data and using it to predict future trends and behaviors; in general, they are intended to answer the question "What may happen?" A variety of statistical methods is engaged to current and historical facts to make predictions about future. In business, a good application example for predictive analytics would be credit scoring that evaluates customers' future creditworthiness based on their credit history, behavior in past loans, spending habits etc.

Predictive analytics have much in common with forecasting, and there's no significant distinction between the two, except that the recent developments in advanced techniques like machine learning allow handling much more varied information for prediction purposes. An example of complex prediction would be CEP (Complex Event Processing) that attempts to recognize complex events in real time, coupled to predictive analytics that supply the expected features of the object of interest, to predict an important event (Fülöp et al., 2012).

Prescriptive Analytics

Prescriptive analytics are intended to answer questions like "What should we do?" or "What is the best course of action under these circumstances?", or, more specifically, e.g. "What would be the best course of action or mix of initiatives against short-term financial gains?" Analysis of available factual data is expected to suggest the possible alternatives of action, or, more commonly, evaluate the known alternatives and select the best one. Modeling, simulation, operations research techniques, multi-criteria decision techniques, scenario generation and evaluation—these are the techniques used by prescriptive analytics. Some of them, like modeling or simulation, have been in use for a considerable time, and for several decades attracted significant attention from researchers and analysts.

7.5.4 Augmented Analytics

Augmented analytics, a rather recent term whose authorship is attributed to Gartner in 2017, are expected to act symbiotically with the human analyst, automating mundane procedures and providing support where possible—mostly in repeating and predictable situations. Augmented analytics seem to have much in common with self-service analytics, hybrid analytics and other types of analytic activities where user activities are balanced with IT support. In addition to this similarity, augmented analytics employ machine learning and natural language processing/generation technology to achieve greater sophistication in their support to the business analyst. This is not exactly a new approach—there have been numerous attempts to put the power of analytical technologies directly into the hands of business users, eliminating the middle persons like data scientists or software consultants. Clearly, an important feature of augmented analytics is the necessity to have a business question beforehand, so it is expected to achieve best results for question-driven, or problem-driven situations, as explained in Chap. 4 of this book.

One important feature of all analytic applications is their rate of repeatability which depends on the repeat rate of situations requiring analysis—decision making, insight development, strategy formation. Repetitive situations create a body of experience and call for tools that are tried, tested and reliable. Situations with low repetition rate tend to have more uncertainty, and exploratory tools like simulation will be more appropriate.

7.5.5   Web Analytics

Although not very original by the nature of used analytical approaches, Web analytics have opened many new possibilities with the spread of Internet technology and e-commerce. Several important forces are at work here:

- For an individual or organization, non-participation on the Web practically means isolation and severe limitation of possibilities;
- Any activity on the Web leaves digital traces that can be collected and processed to look for useful insights.

7.6   Modeling and Simulation

Modeling technologies in the context of business informatics serve an important role in decision making when encountering important problems. The notion of modeling is far wider, however; modeling approaches are used in all creative and problem-solving activities. Models are based on assumptions, serve insight building by managing abstraction, and are inexact reflection of the original.

The range of business models extends from simple to the very complex. Simple models might be used every day—e.g., embedded classifying model used by sales representative and based on a neural network that recognizes an instance of certain client behavior, and recommends promotional action. Complex simulation models might be one-of-a kind and rather expensive to develop—e.g., unique forecasting model to evaluate several market penetration strategies and containing multiple variables, calculations and scenarios. Regardless of their complexity, models have one common important feature—a "what-if" capability to experiment and evaluate scenarios easily and without limits.

By the nature of their components and logic of relations between them, models can be either quantitative or qualitative. Quantitative models handle the measurable

Table 7.3 Categories of models (based on Sharda, Delen, & Turban (2014))

Category	Process and objective	Techniques
Optimization	Find the best solution from a number of alternatives	Decision tables, decision trees, linear and other optimizing models, inventory models, transportation models, critical path models
Simulation	Find a good enough solution or try out several alternatives by computer simulation of objects or processes	Monte-Carlo models, queueing models, system dynamics
Heuristics	Find a good enough solution using rules	Heuristic programming, expert systems
Predictive models	Predict the future developments for a given scenario	Forecasting models, Markov models
Other models	What-if experiments	Financial modeling

features of the objects of interest and use advanced mathematical techniques—quantitative methods, forecasting, simulation. Qualitative models reflect features that are not easily quantifiable, and either attempt to quantify them (e.g., using Likert scales) or attempt to put together meanings and soft characteristics, like customer opinions or organization culture.

Several broad categories of models are briefly described in the Table 7.3 below, based on Sharda, Delen, & Turban (2014).

The principal types of computer models for business have been developed alongside the advent of first computers and developments in computational methods. Although the algorithms used in models do not change much, the software incorporating these algorithms has undergone many phases of sophistication regarding functionality, flexibility and ease of use. The off-the-shelf products for business modeling are quite numerous and more of a niche product than a generic product for mass use. Nevertheless, some better-known modeling software packages that represent the basic types of models are: MatLab (a suite product for an array of modeling applications); GAMS and LINDO for optimization models; AnyLogic, FlexSim, PowerSim for simulation models; project management suites like Microsoft Project utilizing critical path methods (CPM); multicriteria evaluation packages like ExpertChoice or Banxia. Of course, an Excel spreadsheet is and will probably remain an ubiquitous tool for building simple models, and there are simple and inexpensive Excel extensions like Palisade @RISK that are used for quick modeling of repetitive problems. A comprehensive collection of simulation software is provided on the Web site of Capterra—an integrator of research and user reviews on software applications for business (Simulation Software, n.d.).

Summing up, modeling technologies are a powerful tool, whose potential is best utilized in situations that are typical for a known class of models. As such situations usually are not commonplace, modeling technologies can be considered "heavy artillery"—powerful, yet expensive and requiring special competencies. The more recent modes of technology deployment, like SaaS (software-as-a-service) under cloud technologies, have already proved useful in managing ownership costs for expensive modeling software.

7.7   Text Analytics

Text analytics are a relatively new breed of analytical technologies employing computer linguistics with statistics to derive additional information, possessing a higher level of aggregation. This information is expected to be of high quality because of its novelty, relevance and disclosure of previously unknown and potentially valuable relations. These features are quite close to the definitions of the result of data mining: to produce information that is new (previously unknown), useful, not evident, and cannot be extracted by ordinary (simple) means. This resemblance, together with the wide use of statistical methods, has born the term "text mining" that is often used as another term for text analytics, although some differences may be spotted; for example, text analytics includes information retrieval that is outside mining approaches. Hearst (1999) indicated differences between text mining and information retrieval. For information retrieval, the required information (document) is known to exist; a special information need, zooming in on a single specific piece of information, is satisfied by search and retrieval technology. Text mining looks for new information that has not been discovered before.

A distinctive feature of text analytics is the wide use of natural language processing (NLP) technologies that are an important area of artificial intelligence.

Several well-known examples of text analytics:

- Use in security applications: ECHELON surveillance system has been the first of a kind; later, EUROPOL, FBI, CIA and other security agencies have developed systems to monitor communication flows of all modes—telephone, emails, other transmissions to track threats like terrorist activities or transnational crime.
- A widely quoted example of text analytics application is the performance of IBM Watson system at Jeopardy! game, where the system came up winning against top human performers. To achieve this, a Watson subsystem with massively parallel architecture, called DeepQA, has been developed using more than 100 different text mining techniques.
- Patent intelligence—many companies in mostly technology industries, but not limited to them, have been scanning patent collections for many years to detect innovations, track competitor moves, identify and recruit the creative talents behind fresh patents, identify patent infringements and do much more to maintain awareness in the competitive landscape.

Some issues of text analytics have been covered in Chap. 5, mostly related to information extraction. In addition to information extraction, text analytics use a multitude of other methods, such as (Sharda, Delen, & Turban, 2014): topic tracking, categorization, clustering or concept linking. Each application of these methods has to produce a specific structure that discovers previously unknown and useful

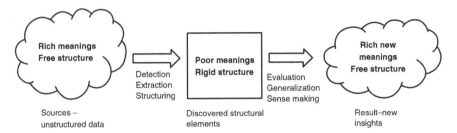

Fig. 7.3 Stages of a generic text analytic process

information as a result. So the rough stages of a generic text analytics process can be defined as follows (Fig. 7.3):

- Unstructured (source data; rich meanings and free, flexible structure)
- Structure (graphs, table, WordCloud etc.; poor meanings and rigid structure)
- Unstructured (insights; rich new meaning and little need for structure).

The goal of the process above is to discover new useful information and insights through introduction of structure—discovered elements, relations, rules. Text analysis finds elements of structure and measures their statistics—e.g., defines clusters who might or might not be considered new information, but in any case they become an object of further analysis. There is a risk that in the process of structuring, which mostly regards lexical and syntactical levels, some meanings at the semantic level will be lost, so the area of semantic level in text analytics has attracted growing research interest (Sinoara, Antunes, & Rezende, 2017).

The methods of text analytics range from simple, e.g., WordCloud, to rather sophisticated like IBM Watson DeepQA. An example of WordlCloud use is shown in Fig. 7.4.

According to the producers of one of the more prominent text analytics products, Megaputer PolyAnalyst, a market product for text analytics should contain all or some of the below functions:

- A variety of data sources and acceptable formats;
- Dictionaries, thesauri and their editors;
- Tools for defining patterns;
- Terms extraction, linking and clustering;
- Document clustering;
- Text dimensions and use of OLAP principles;
- Visualization and dashboards.

Domain dependence. Text analytics may or may not depend upon the subject area, and two approaches dominate (Cohen & Hunter, 2008; Tan, 1999):

Fig. 7.4 An example of word cloud (https://getthematic.com/insights/the-5-major-faults-of-word-clouds-and-how-they-harm-your-insights/)

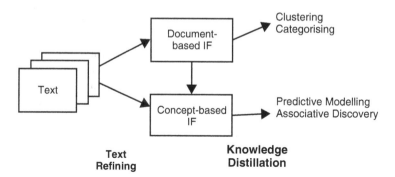

Fig. 7.5 A domain-dependent framework for text mining (Tan, 1999)

- Linguistic, or rule-based approaches—depend on the subject area and uses some body of knowledge or a collection of area-specific terms;
- Statistical approaches—independent of the subject area; use statistical analysis and build classifiers to detect useful patterns.

Tan (1999) has proposed a framework for text mining that comprises two components, or phases, and illustrates domain dependency (Fig. 7.5):

- Text refining, transforming free-text documents into an intermediate form, and
- Knowledge distillation, deducing patterns or knowledge from the intermediate form.

The first phase can convert free-text documents into two types of intermediate forms: document-based or concept-based. In the first case, each entity represents a document, and usually is domain-dependent. In the second case, each entity represents a concept of interest in a given domain, and is domain-independent.

Regarding domain dependence, it is interesting to note that text mining, being in many cases rather dependent on the subject area, often has been driven by the area professionals, rather than by text mining (technology) specialists. According to Cohen and Hunter (2008), the text mining systems built and deployed in the field of bioscience, and featuring high usage rates, have been built by bio-scientists.

Use of elements of data analytics. As the process of text analytics develops certain data structures, it borrows some principles from the processes of analytics of structured data. Park and Song (2011) describe an approach to text analytics that they call a *Text OLAP* approach. Similar to OLAP use with data bases and data warehouses, the proposal relates a document to a line in fact table, and defined keywords serve as dimensions. Another such borrowed principle is a so-called *textual ETL* that has a role in the technology called textual disambiguation and developed by Bill Inmon in 2012. Textual disambiguation applies context to raw text, and reformats the raw text with related context into a standard data base format. Being structured in this way, text can easily be accessed and analyzed by standard business intelligence technology. This approach can also be seen as a simple attempt to introduce structure and manage context.

Sentiment analysis. The growing volumes of user-generated content contain opinions, evaluations and emotions that may regard an important object and contain value for the interested parties. This interest brings up text analysis issues like sentiment polarity and ranking: while clearly negative and clearly positive responses are easy to detect and count, the ranks in between form an interim zone where polarity is not so simple to detect. Li and Li (2013), researching Twitter messages and microblogs, pointed to complications caused by rapid development of new vocabulary in tweets—abbreviations, emoticons etc. Hasan and Curry (2014), discussing event processing, describe folksonomies—bottom-up social tagging that produces user-generated taxonomies instead of those imposed top-down in advance.

Software products. The most known products for text analytics and mining currently are: Microsoft Text Analytics, IBM Watson Explorer, Amazon Comprehend, SAS Text Miner, Megaputer PolyAnalyst, MonkeyLearn and others. We can also mention qualitative analysis packages NVivo, Atlas.ti; although they are not exactly text analytics software, they employ advanced information management techniques that analyze unstructured information—interviews, transcripts and the like.

7.8   Presentation and Visualization

Data visualization defines a task of designing a graphical representation of data to support its interpretation by users. Data presented as text or structured collections reflect the exact facts, yet do not allow easy detection of trends and patterns. On the other hand, a graph alone, however picturesque, does not reflect the level of detail that is represented by data itself, so the two views (rough data and graph) usually are complementary.

Although charts and graphs depicting business data and information have been in existence for a long time, the spread of computer technology, especially personal computing devices, has made visualization easy for virtually everyone. Historically, spreadsheet packages were the most popular platform to introduce simple charting and graphing tools. With growing needs for visual impact, numerous specialized visualization tools appeared on the market, offered as independent software products or parts of larger information systems. For modern business intelligence, data visualization has become the de facto standard, and virtually all software attributed to BI or business analytics has strong data visualization functionality. The role of visualization tools has been reinforced lately by the coming of Big Data and the needs to uncover patterns, trends and anomalies hidden in data collections. Having accumulated huge collections of data, business users need graphic options to get fast and convincing analytic views of their data.

Analytical approaches that are used in today's business have their roots in other areas, e.g., scientific research; and BI need for visualization emerges from such roots. The historical roots of visualization are much deeper, and there are rather interesting examples of important visualizations that were instrumental in reinforcing the story or even changing the course of events. Three of such examples are presented below (Figs. 7.6, 7.7, and 7.8).

The famous drawing of slave bodies packed into the slave ship "Brookes", published by the abolitionist society of Plymouth, Massachusetts, helped gaining support for the abolitionist movement to end slavery in the United States.

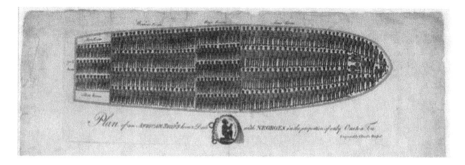

Fig. 7.6 Drawing of slave ship 'Brooks' (Bristol Museums, Galleries and Archives)

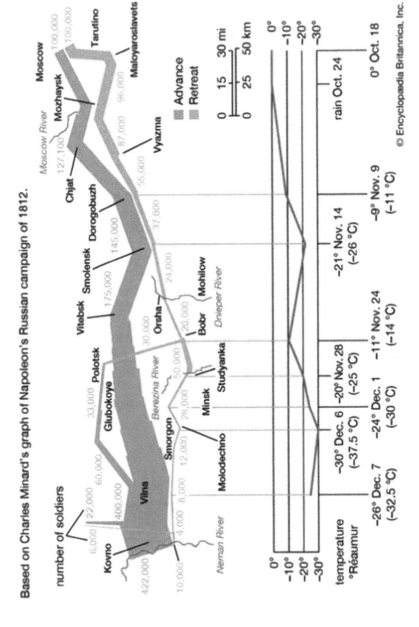

Fig. 7.7 Napoleon's Russian Campaign Infographic (Rendgen, 2018)

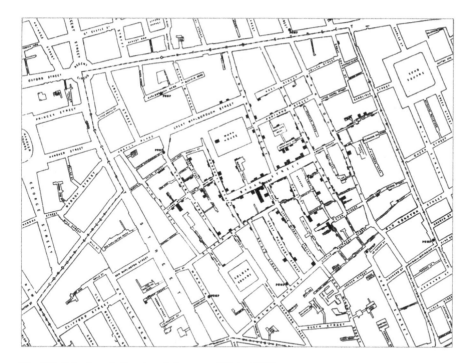

Fig. 7.8 John Snow's London cholera map (Klein, 2014)

The drawing of Napoleon's army Russian campaign route of 1812 rather clearly depicts the story using several dimensions: geographic directions that are location-correct; number of soldiers depicted by the thickness of the direction line; and weather temperatures (added later).

Another famous historical visualization, John Snow's London cholera map of 1854, eventually helped to locate the infection epicenter—the infected water pump on Broad Street, which happened to be almost the exact center for the cluster of dark (infected) areas on the map. Of the more recent famous visualization examples, we should mention the Trendalyzer visualization software, developed by Hans Rosling's Gapminder Foundation (Rosling, 2006).

One of the most common visualization formats in BI environments is dashboards, together with mashups mentioned in Chap. 5. Dashboards are intended to reflect the most important information, often of diagnostic nature in some form of KPI (Key Performance Indicators) in a condensed, presumably easy to use format.

The emergence of dashboards in business intelligence is logical and probably will have a solid life cycle, because it provides an opportunity for business users to have critical information at their fingertips in a condensed form. First embryo dashboards emerged with initial designs for executive information systems (EIS) in the 80s, just to be put on hold because of inability to provide reliable information "feed"; and reborn again in the end of XX century. Visualization technology will undoubtedly provide new opportunities and features which will be weighed against the same principles of data relevance, view composition, attention management and simply good taste. A body of experience has been already accumulated in dashboard use, and many professional sources name some common good or not-so-good features:

Good Features:

- Simplicity—reasonable number of data elements (widgets); uncluttered view;
- Interactivity—e.g., sliders setting a range, or other user-friendly manipulation like drill-down;
- Color consistency—subdued for general view; accentuated to attract attention;
- Geolocation display where appropriate (location-dependent);
- The northwest corner is the point of prime interest, and the most important information should be placed there;
- If several data series use same scale (e.g., time), it makes sense to justify data along the same axis.

Bad Features:

- Low quality, unreliable data;
- Overload with content; too many indicators;
- Small and hard-to-read charts;
- Color abundance; dark or black backgrounds are not the best solution;
- Chart choice—if the chosen chart type doesn't tell much, graph type should be changed if the data is important, or the graph should be removed altogether;
- Lack of context if some key data around the graph is missing;
- Too many data points for this type of graph;
- Over-decorated settings that do not add to sense making;
- Use of pie charts is limited; even more so for gauge charts;
- Too large data precision—too many digits;
- Unordered (unsorted) bar graphs, if not representing time series;

Although dashboard is a relatively simple concept, research in information presentation has brought up issues directly related to dashboards. Wong, Seidler, Kodagoda, and Rooney (2018) have defined visual persistence as a rate of matching data view to analytical functions: "Persistence occurs when data and the state of one's analysis and reasoning are made visible and remain in view. It enables the analyst to off-load memory challenging activities such as recalling facts to the interface." Evidently, dashboards rather conform to the requirements of persistence. The notion of dataspaces (Franklin, Halevy, & Maier, 2005) is quite close to dashboards as well. One of the pioneering visualization researchers, Ben Shneiderman (1994), had proposed dynamic visual querying that preceded flexible dashboards by several years.

The most common graph or chart types used for business data are presented in Table 7.4.

Table 7.4 Common business graph types

Graph group	Graph type
Single series	Pie, donut
Multiple series	Line, bar, stacked bar, area, stacked area, radial
Combined	Combo chart, box plot
Cartesian	X-Y, scatter plot, bubble graph
Time stages	Funnel, waterfall, Gantt chart

Source: author

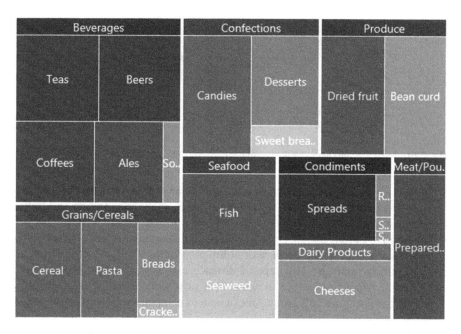

Fig. 7.9 TreeMap graph (https://www.grapecity.com/wijmo/docs/Topics/Chart/Advanced/SpecialCharts/TreeMap)

Other important and more advanced types of visualization include TreeMaps, Surface graphs, Infographics, and graph analytics. Some examples are given below (Figs. 7.9, 7.10, and 7.11).

Graph analytics. Graph analytics, being a relatively new mode of visualization, merit a more detailed look. Based on graphs and node connectivity, they proved rather useful in social relations. According to Ferguson (2016), four widely used types of graph analytics include path analysis, connectivity analysis, community analysis and centrality analysis:

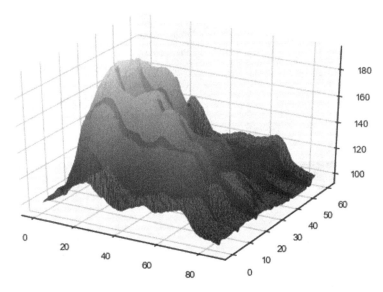

Fig. 7.10 Surface graph (https://python-graph-gallery.com/371-surface-plot/)

- **Path analysis:** This type of analysis determines the shortest distance between two nodes in a graph, and is useful in supply chains, route optimization and traffic optimization.
- **Connectivity analysis:** This type of analysis determines weaknesses in networks such as communications network or utility power grid.
- **Community analysis:** This type of distance and density–based analysis finds groups of interacting people in some kind of a social network, and identifies whether the network will grow. An example of community network is presented below in Fig. 7.12.
- **Centrality analysis:** This type of analysis enables identifying relevancy to find the most influential people in a social network, or to find most highly accessed web pages—such as by using the PageRank algorithm.

Graph analytics process two types of objects—nodes, or vertices; and connections, or edges. Nodes may be people, or people groups, or network nodes, or any other nodes. Edges represent relationships such as emails, payments, phone calls, physical trips, social networking etc. Numerous use cases for graph analytics exist (Ferguson, 2016)

- Detecting money laundering
- Spotting fraud in government, banking, insurance
- Preventing crime and terrorism
- Applying influencer analysis in social network communities
- Evaluating grid and network quality of service such as identifying weaknesses in power grids, water grids, communication and transportation networks

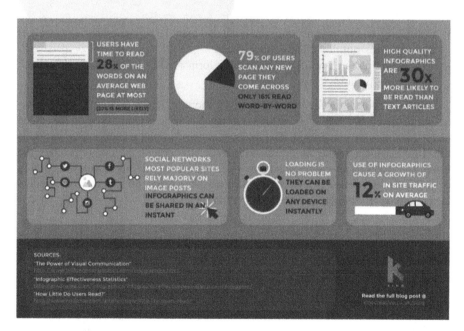

Fig. 7.11 An example of infographics (https://visual.ly/community/Infographics/technology/infographics-benefits-their-use-online?utm_source=visually_embed)

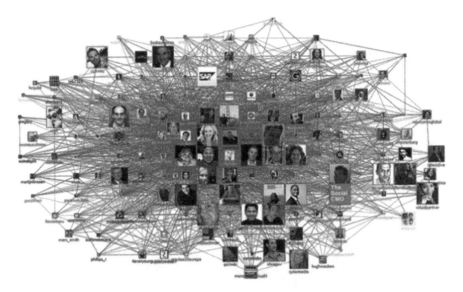

Fig. 7.12 An example of graph analytics on a community network (Ferguson, 2016)

- Optimizing routes in transportation, retail and manufacturing industries as well as for supply distribution chains and logistics
- Conducting research in life sciences (bioinformatics) including medical research and disease pathologies

A principal medium for visualization is a monitor screen, which has its limitations: the screen size is limited, so the importance of what is going to be on the screen has to be seriously considered. The other, to author's opinion, serious limitation is the 2-dimensional mode of presented view. Although it perfectly matches the visualization on paper media, many new possibilities would open with the coming of 3-D display mode. However, at the time of writing this book, 3-D display technology is still rather far from being commonplace.

7.9   The Role of AI

The current resurgence of artificial intelligence (AI) techniques has its roots in early AI—expert systems, NLP, but mostly neural networks (NN) that have an ability to learn, or improve their performance by adjusting their prediction mechanism by the results of statistical analysis of factual data. This ability is currently widely discussed as machine learning (ML). Neural networks have been around for a substantial time, starting with work by McCulloch and Pitts in 1943, and followed by a string of many other, sometimes fragmented, developments. A recent rise of interest in machine learning techniques has added momentum to research in NN field. As the flexibility and analytical power of NN grew, application areas expanded from simple tasks like character recognition to complex tasks like trading in stock markets. By their learning mode, NN divide into two groups:

- *Supervised learning*—a network is trained with known pairs of input-output, or known examples where a set or several sets of input signals should result in a certain output. Classification is used for category results, and regression—for numeric results. The most common algorithms are linear regression for regressions, and random forest and support vector machines for classification problems.
- *Unsupervised learning*—a network is trained with data that has no known outcomes, and is used for discovery of clusters, anomalies, associations. Apriori algorithms are used for association rule learning, and k-means algorithms for clustering.

Between the two modes of learning, medium ground is possible where only several of the possible outcomes can be named, or labelled. Many real-world classification cases fall into this category.

The field of NN and ML contains many more specific terms that define one or other type of structure or behavior: feed-forward, or forward propagating; back propagating; recursive; recurrent; convolutional; Kohonen networks; Hopfield networks and others. These are key topics when an interest in machine learning arises; and for more detail, there is ample literature on NN and ML; e.g., Aggarwal (2018); Domingos (2018) or Goodfellow, Bengio, and Courville (2016).

When discussing the prospective areas of AI application, many professional and market research sources currently note the spread of AI into clerical and managerial functions, where there are repeating sets of conditions and classifier algorithms. This inevitable automation of repeating tasks is sometimes called process intelligence, and includes customer service robots, question-answering applications, inquiry processing and many others. Such repeating tasks of not-so-complex nature seem to have a huge cost-saving potential. The more complex tasks need more powerful approaches, and one of such approaches is called deep learning.

Deep learning. Deep learning approaches are based on several algorithms: deep neural networks (the dominating approach), and random decision forests that use a set of decision trees and average their output. Deep neural networks have multiple hidden layers and can model complex non-linear relationships. Deep learning applications handle complex tasks: computer vision, speech generation, autonomous vehicles.

Augmented AI. Augmented intelligence is a mode of AI application that encompasses joint analytical competencies of human users and AI capabilities. According to Domingos (2012), data alone is not enough to leverage ML strengths; every ML situation has to embody some knowledge or assumptions beyond the data to act as a starting point—"it (ML) can't get something from nothing".

Deloitte Analytics senior adviser Thomas H. Davenport (Mittal, Kuder, & Hans, 2019) describes three stages in utilization of artificial intelligence:

- In the first stage, *assisted intelligence*, companies harness large-scale data programs and science-based approaches to make data-driven business decisions;
- Currently, companies are already working toward the next stage—*augmented intelligence*—in which ML capabilities layered on top of existing information management systems work to augment human analytical competencies;
- In the coming years, more companies will progress toward *autonomous intelligence*, the third AI utilization stage, in which processes are digitized and automated to a degree whereby machines, bots, and systems can directly act upon intelligence derived from them.

Whereas the stage of autonomous intelligence sounds somewhat optimistic, the second stage of augmented analytics appears to be rather reasonable, given the recent research interest in hybrid or augmented systems that utilize joint human-technology competencies in a best way. The drive of AI research to replicate the human mind is perfectly understandable; however, it leads to extremely complex products whose behavior might be very complicated to understand and predict. To author's opinion, if, alongside with the quest to recreate a human mind, more effort would be directed at the hybrid, or augmented intelligence, AI approaches would only benefit from better understanding of how a human-machine tandem can get smarter.

7.10 Communication and Collaboration Platforms

Since the breakthrough of mid-eighties in the area of group decision support, primarily by the researchers of University of Arizona—D. George. D. Vogel, J. Nunamaker and others—a distinct class of collaboration technologies has been defined, together with a then-new class of decision support systems—group decision support systems. Decisions often have to be taken in ad hoc and collaborative manner (Berthold et al., 2010), and combination of intelligence functions with collaborative platforms can dramatically improve the quality of decision making.

The blend of business intelligence functions with collaboration platforms, including group support and social platforms, has been labelled collaborative BI. The coming of social platforms into the workplace has created new possibilities for collaboration. Regarding BI, numerous sources have stressed the importance of horizontal cross-functional exchanges between analysts (Grublješič & Jaklič, 2015; Presthus, 2014; Shah, Horne, & Capella, 2012). The importance of the collaborative functions in BI is obvious for many reasons:

- The width of situations requiring insights for decision making more often than not crosses the boundaries of a single function or organizational unit;
- The shared data models and analytical tools promote unified understanding of business context and prevention of incompatible data, duplicated efforts or erratic models;
- The shared facts, assumptions, insights and expertise to cover gaps provide significant insurance against decision errors;

– The experience, gained in previous analyses and decision making, should be accumulated, organized and shared.

The actual technology for collaborative BI does not have to be significantly different from collaborative technologies in general, so a variety of technical platforms is applicable here, from internal social networks and Intranets to custom-designed collaboration and sharing platforms for specific activities. Gewehr, Gebel-Sauer, and Schubert (2017) have proposed a social network of business objects—an environment uniting participants, snippets of information and simple interface that is expected to serve non-standard, ad-hoc intelligence needs. Matzler, Strobl, & Bailom (2016) described the use of a competitive intelligence integration platform, called "competitor Wiki" that supported integration of competitor information and the quality of resulting insights. Some proprietary platforms for collaborative BI have been patented (e.g., Dayal et al. (2010)).

It has to be noted that the existence of a simple, shared and open IT platform for collaborative BI not only assists in solving inconsistencies in insight development, but also significantly supports the development of BI culture (Skyrius, Nemitko, & Taločka, 2018) by supporting formation and strengthening of BI communities. The issues of BI culture are discussed in more detail in Chap. 9 of this book.

7.11   BI Technology Deployment Issues

As BI adoption in an organization grows, the variety of options for BI technology deployment grows as well. While attracting modest research interest, the issues of BI deployment are nevertheless rather important for business managers and IT professionals who deal with projects of BI implementation. A thorough cross-projection of BI requirements into deployment options, centering on data deployment issues, is given in (Zangaglia, 2006). Boyer, Frank, Green, Harris, and Van De Vanter (2010) has presented four BI deployment paradigms along the dimension of flexibility and control on one side, and time-to-value and possible savings on the other:

• Enterprise software deployment, or use of enterprise software as a base for BI seeding—this option provides most control and flexibility;
• Virtualization, joining available array of technology resources under umbrella metasystem;
• Optimized business systems with gradual implementation of packaged turnkey BI products;
• Cloud computing, offering BI functionality under SaaS (Software-as-a-service) format—this option offers shortest time-to-value and possible lowest cost options.

It is hardly possible to define a universal set of criteria for technology decisions; yet, considering BI dimensions mentioned earlier in Chap. 4, several common considerations do exist and are discussed in more detail below.

Deployment scale. The planned BI system can be implemented at once in its full width, or undergo a phased deployment, usually starting with so-called pain points where fast returns can be expected and demonstrated. Many practical and professional sources recommend phased deployment as less risky for BI projects, whose activities are far more turbulent than those of regular business systems, and stable maturity is virtually impossible to achieve.

Organization of data resources. The choice is between dedicated data warehouse and virtual data integration. The arguments for selecting one or another solution have been discussed in detail in the beginning of this chapter—arguments like flexibility and adaptability versus efficiency and reliability are applied in many types of systems, including solutions for the setup of data resources.

The width of function set. This set can be company-wide or narrow project for a single function. As noted above regarding deployment scale, it makes sense to start with a narrow application area whose benefit potential is easily defined, and expand from there.

The number of standards. The definition of the set of standards is both art and science, as there is no exact method to determine the best set for a given implementation project in a given organization. It is obvious that too few standards will create significant problems in integration of data and functions, while too many standards will limit flexibility and future expansion.

On-site deployment versus cloud/SaaS. There are several points to consider when deciding for or against cloud option. Traditional on-premises solutions are the ones usually recommended for BI projects, firstly because of increased risks for important decisions because of system inaccessibility or compromising sensitive BI information. The additional points for on-site case may be less complicated integration of BI and other heterogeneous internal information systems, and tradition to rely on own capacity. Cloud solutions might be recommended in the case of need to temporarily use specific advanced technologies that would be too expensive to have on-site; or when IT resources and budgets are strained. Cloud solutions are considered for specialized functions such as website analytics, text mining or speech analytics. Organizations can sample using cloud-based BI technology while waiting for the installation of on-site solutions.

Standard off-the-shelf technology versus custom-made or open source. Standard off-the-shelf BI technology is a common case for on-site installations. Custom-made BI systems are a much more rare case. Although open source is an attractive option because it doesn't incur costs anything to obtain, installing and using it requires expert technical knowledge. Open source or custom-made systems are also an option when BI needs are likely to require customization in the future.

Mobility. The mobility of BI function is not a part of opposing dimension—it is just a mobile image of BI system in use. Some BI technology vendors have integrated mobile capabilities into their existing products, while other solutions require additional resources for mobile support. A push or pull technology can be chosen to determine whether the most current data will be broadcast to the mobile devices, or users will have to pull the data they need. The obvious benefit of mobile BI is access from anywhere, which might be useful making decisions away from head office. With mobile devices, security is a key consideration; its level is determined by the sensitivity of information that can be accessed in the BI system.

References

Adriaans, P., & Zantinge, D. (1996). *Data mining*. Harlow: Addison-Wesley Longman.

Advanced Analytics. (n.d.). *Gartner glossary*. Retrieved September 30, 2019, from https://www. gartner.com/en/information-technology/glossary/advanced-analytics.

Aggarwal, C. C. (2018). *Neural networks and deep learning*. Cham, Switzerland: Springer International.

Agrawal, R., & Srikant, R. (1994). Fast algorithms for mining association rules. In *Proceedings of the 20th international conference on very large data bases* (pp. 487–499). Chile: Santiago.

Arena, P., Rhody, S., & Stavrianos, M. (2005). Using advanced analytics to drive better business decisions. *Information Management*, April 1. Retrieved September 15, 2007, from www. information-management.com/issues/20050401/1023895-1.html?zkPrintable=true.

Arnott, D., & Pervan, G. (2014). A critical analysis of decision support systems research revisited: The rise of design science. *Journal of Information Technology, 29*, 269–293.

Berry, M. J. A., & Linoff, G. S. (2004). *Data mining techniques* (2nd ed.). Indianapolis, IN: Wiley Publishing.

Berthold, H., Roesch, P., Zoller, S., Wortmann, F., Careninin, A., Campbell, S., Bison, P., & Strohmaier, F. (2010). An architecture for Ad-hoc and collaborative business intelligence. In *EDBT 2010: Proceedings of the 2010 EDBT/ICDT Workshops, March 22–26, 2010* (pp. 1–6). Lausanne, Switzerland: EDBT.

Biscobing, J. (2019). *OLAP (Online Analytical Processing). Interactive*: Retrieved September 30, 2019, from https://searchdatamanagement.techtarget.com/definition/OLAP.

Boyer, J., Frank, B., Green, B., Harris, T., & Van De Vanter, K. (2010). *Business intelligence strategy: A practical guide for achieving BI excellence*. Ketchum, ID: MC Press Online.

Breiman, L., Friedman, J. H., Olshen, R. A., & Stone, C. J. (1984). *Classification and regression trees*. Boca Raton, FL: Chapman & Hall.

Cohen, K. B., & Hunter, L. (2008). Getting started in text mining. *PLoS Computational Biology, 4* (1), e20.

Dayal, U., Vennelakanti, R., Sharma, R. K., Gastellanos, M. G., Hao, M. C., Patel, C., Bellad, S. S., & Gupta, M. (2010). *US Patent No. US20100325206A1*. Washington, DC, USA: US Patent and Trademark Office.

Domingos, P. (2012). A few useful things to know about machine learning. *Communications of the ACM, 55*(10), 78–87.

Domingos, P. (2018). *The master algorithm: How the quest for the ultimate learning machine will remake our world*. New York, NY: Basic Books.

Elliott, T. (2004). *Choosing a business intelligence standard. Business objects white paper*.

Ferguson, M. (2016). *What is graph analytics?* Interactive: https://www.ibmbigdatahub.com/blog/ what-graph-analytics.

Franklin, M., Halevy, A., & Maier, D. (2005). From databases to dataspaces: a new abstraction for information management. *SIGMOD Record, 34*(4), 27–33.

Fülöp, L. J., Beszédes, Á., Tóth, G., Demeter, H., Vidács, L., & Farkas, L. (2012). Predictive complex event processing: A conceptual framework for combining complex vent processing and predictive analytics. *BCI '12: Proceedings of the Fifth Balkan Conference in Informatics*, September 2012, Novi Sad, Serbia, 26–31.

Gewehr, B., Gebel-Sauer, B., & Schubert, P. (2017). Social Network of Business Objects (SoNBO): An innovative concept for information integration in enterprise systems. *Proceedings of CENTERIS/ProjMAN/HCist 2017*, Barcelona, Spain, pp. 904–912.

Goodfellow, I., Bengio, Y., & Courville, A. (2016). *Deep learning*. New York: MIT Press.

Grublješič, T., & Jaklič, J. (2015). Business intelligence acceptance: The prominence of organizational factors. *Information Systems Management, 32*, 299–315.

Guha, S., Rajeev Rastogi, R., & Shim, K. (2000). ROCK: A robust clustering algorithm for categorical attributes. *Information Systems, 25*(5), 345–366.

Hasan, S., & Curry, E. (2014). Thematic event processing. In *ACM/IFIP/USENIX Middleware conference 2014* (pp. 109–120). New York, NY: ACM Press.

Hearst, M. A. (1999). Untangling text data mining. *ACL '99: Proceedings of the 37th annual meeting of the Association for computational linguistics on computational linguistics*, June 1999, pp. 3–10.

Hidber, C. (1998). Online association rule mining. *SIGMOD '99: Proceedings of the 1999 ACM SIGMOD International Conference on Management of Data*, June 1999, Philadelphia, PA, USA. pp. 145–156.

Inmon, W. (1992). *Building the data warehouse*. Indianapolis, IN: Wiley.

Jeusfeld, M. A., & Thoun, S. (2016). Key performance indicators in data warehouses. In E. Zimányi & A. Abelló (Eds.), *Business intelligence—5th European Summer School, eBISS 2015*, Barcelona, Spain, July 5–10, 2015, Tutorial Notes, pp. 111–129.

Klein, G. (2014). *Seeing what other's don't*. New York, NY: Public Affairs.

Kranc, M. (2017). *Is data vault a good choice for your organization?* Retrieved September 21, 2019, from https://insidebigdata.com/2017/07/28/data-vault-modeling-good-choice-organization/.

Li, Y.-M., & Li, T. Y. (2013). Deriving market intelligence from microblogs. *Decision Support Systems, 55*, 206–217.

Linstedt, D. (2002). *Data vault series 1—Data vault overview*. Retrieved September 23, 2019, from https://tdan.com/data-vault-series-1-data-vault-overview/5054.

March, S. T., & Hevner, A. (2007). Integrated decision support systems: A data warehousing perspective. *Decision Support Systems, 43*, 1031–1043.

Matzler, K., Strobl, A., & Bailom, F. (2016). Leadership and the wisdom of crowds: how to tap into the collective intelligence of an organization. *Strategy & Leadership, 44*(1), 30–35.

Mittal, N., Kuder, D., & Hans, S. (2019). *AI-fueled organizations: Reaching AI's full potential in the enterprise. Deloitte Insights*. Retrieved June 15, 2019, from https://www2.deloitte.com/us/en/insights/focus/tech-trends/2019/driving-ai-potential-organizations.html.

Park, B.-K., & Song, I.-Y. (2011). Towards total business intelligence incorporating structured and unstructured data. *BEWEB 2011*, March 25, 2011, Uppsala, Sweden.

Presthus, W. (2014). Breakfast at Tiffany's: The study of a successful business intelligence solution as an information infrastructure. *Twenty Second European Conference on Information Systems* (ECIS), Tel Aviv 2014.

Rangarajan, S. (2016). *Data warehouse design—Inmon versus Kimball*. Retrieved September 23, 2019, from https://tdan.com/data-warehouse-design-inmon-versus-kimball/20300.

Rendgen, S. (2018). *The Minard system: The complete statistical graphics of Charles-Joseph Minard, from the collection of the École Nationale des Ponts et Chaussées*. Princeton: Architectural Press.

Rosling, H. (2006). *The best stats you've ever seen. TED talk*. Retrieved September 30, 2019, from https://www.ted.com/talks/hans_rosling_the_best_stats_you_ve_ever_seen?language=en#t-381433.

Shah, S., Horne, A., & Capella, J. (2012). Good data won't guarantee good decisions. *Harvard Business Review*, April 2012.

Sharda, R., Delen, T., & Turban, E. (2014). *Business Intelligence: A Managerial Perspective on Analytics*. Harlow, UK: Pearson Education Limited.

Simulation Software. (n.d.). Retrieved December 15, 2019, from https://www.capterra.com/simulation-software/.

Sinoara, R. A., Antunes, J., & Rezende, S. O. (2017). Text mining and semantics: A systematic mapping study. *Journal of the Brazilian Computer Society, 23*, 9.

Skiena, S. (2010). *The algorithm design manual* (2nd ed.). New York: Springer Science+Business Media.

Skyrius, R., Nemitko, S., & Taločka, G. (2018). The emerging role of business intelligence culture. *Information Research, 23*(4), 806. Retrieved September 30, 2019, from http://InformationR.net/ir/23-4/paper806.html.

Tan, A.-H. (1999). Text mining: The state of the art and the challenges. *Proceedings of the PAKDD 1999 Workshop on Knowledge Discovery from Advanced Databases.*

Tront, R., & Hoffman, S. (2011). Pervasive business intelligence and the realities of excel. *Business Intelligence Journal, 16*(4), 8–14.

Verhoef, P., Kooge, E., & Walk, N. (2016). *Creating value with big data analytics.* Abingdon, UK: Routledge.

Wong, W. B. L., Seidler, P., Kodagoda, N., & Rooney, C. (2018). Supporting variability in criminal intelligence analysis: From expert intuition to critical and rigorous analysis. In G. Leventakis & M. R. Haberfeld (Eds.), *Societal implications of community-oriented policing and technology.* New York: SpringerBriefs in Criminology. https://doi.org/10.1007/978-3-319-89297-9_1.

Wu, X., Kumar, V., Ross Quinlan, J., et al. (2008). Top 10 algorithms in data mining. *Knowledge and Information Systems, 14*, 1–37. https://doi.org/10.1007/s10115-007-0114-2.

Zangaglia, P. (2006). Business intelligence deployment strategies: A pragmatic pattern-based approach. *Business Intelligence Journal, 11*(3), 52–63.

Zhang, G. P. (2008). Neural Networks for data mining. In O. Maimon & L. Rokach (Eds.), *Soft computing for knowledge discovery and data mining.* Boston, MA: Springer.

Zhang, T., Ramakrishnan, R., & Linvy, M. (1996). BIRCH: An Efficient Data Clustering Method for Large Databases. *Proceedings of 1996 ACM-SIGMOD, International Conference on Management of Data*, Montreal, Quebec.

Chapter 8
Business Intelligence Maturity and Agility

8.1 The Definitions and Focus of Business Intelligence Maturity

The discussion of the object of BI maturity might seem outdated—the most important maturity models came up in the years of 2005–2010, and the discussion on their usefulness declined shortly afterwards.

To author's opinion, although past its peak, BI maturity is still a relevant issue because of several reasons:

- Non-diminishing importance of BI and innovative new technologies coming under BI umbrella;
- Need to understand the dynamics of their adoption;
- Relation of the notion of maturity to many important and under-researched BI features, like agility or culture.

The perception of maturity is based on the current experience and understanding of a certain phenomenon, so maturity takes time to accumulate the experience. The current spread of BI applications has created a significant body of experience on BI adoption and implementation. Apart from it, maturity usually encompasses expectations, so to define the future steps is to some extent a subjective measure.

What is meant by immature or mature BI? Most BI maturity concepts easily define the immature part of the maturity cycle—generally it covers the initial phases that have to cover the gap from chaotic BI attempts to managed processes; an immature BI is still under acceptance. According to Gifford (2011), immature BI is when "... BI solutions are piecemeal, inconsistent, and limited in scope. Each MOM (Manufacturing Operations Management), ERP and SCM application creates its own BI capability as well as servicing transactional requirements, without leveraging BI capabilities elsewhere in the enterprise. As a result, business functions depending on effective use of enterprise wide information produce suboptimal decisions limited in scope". There is a recognized gap between the maturity of BI

© Springer Nature Switzerland AG 2021
R. Skyrius, *Business Intelligence*, Progress in IS,
https://doi.org/10.1007/978-3-030-67032-0_8

technology and that of management capabilities: IT technology seems to be mature (Hallikainen, Merisalo-Rantanen, & Marjanovic, 2012), while management capabilities and human issues seem to be far from maturity.

Maturity of a certain phenomenon—technology, innovation, system type—may be related to Gartner's hype curve—the hunch of expectations and disappointment might be of different sizes for different phenomena, but gradually real value gets recognized and essentially sets the duration of the life-cycle for a certain innovation.

The traits of mature BI seem to be less agreed upon, if compared to immature part. An unifying feature of known maturity approaches and models is the expression of expectations in some, often idealistic form. There are examples where in a single organization some BI aspects are more mature than others (Skyrius & Rėbždaitė, 2011). To address this unevenness, some models contain a matrix-like structure with BI asset areas projected against maturity stages (Cates, Gill, & Zeituny, 2005; Eckerson, 2007; Tan, Sim, & Yeoh, 2011). Certain BI tools, like data warehouses, may be considered mature because they are tried and tested many times, but their adoption in a given organization may stumble, because the organization itself is BI-immature in the terms of maturity in information management, analytics and decision making. BI maturity is based not so much on the maturity of infrastructure, as on the maturity of extracting appropriate support from this infrastructure.

It might seem appropriate to relate together organizational maturity and BI maturity, or, more consistently, organizational, informational and BI maturity, assuming that for BI to become mature, a corresponding level of organizational and informational maturity is required. Some organizations may be rather mature and measure their experience in hundreds of years—e.g., universities or banks. However, their information maturity and BI maturity may experience problems, especially for intelligence where agility and flexibility are required.

8.2 Business Intelligence Maturity Models, Comparison and Limitations

Many sources present BI as a nearly mature phenomenon, but a closer look at the business intelligence maturity models (BIMM) raises some doubts about ultimate maturity being the case. The known BI maturity models roughly follow the guidelines set by Richard Nolan's stages-of-growth model (Nolan, 1973) and Watts Humphrey's Capability Maturity Model (Humphrey, 1988). Since then, the notion of maturity has been used in many domains concerned with innovations—e.g., business process maturity (Van Looy, Poels, & Snoeck, 2016), project management maturity (Crawford, 2006), project management in law firms (Gottschalk, 2002), or any domain (de Bruin, Freeze, Kulkarni, & Rosemann, 2005). The object of maturity may differ, but the principles remain—it defines stages of how a certain innovation proves its value and receives acceptance.

Table 8.1 Reflection of BI dimensions in maturity models

Dimension	Immature BI	Mature BI
Internal-external orientation	Internal orientation dominates	Inclusion and integration of internal and external information
Centralization-decentralization	Scattered and decentralized	Centralized infrastructure and decentralized functions
Simple to complex questions	Mostly simple; complex questions need manual or custom approaches	Covers the range from simple to complex, with infrastructure equally easy to use for both types
Functional scope narrow to wide	Narrow and patchy	Diverse and rich; encompasses all activities
Automation level	Manual or one-time solutions	Automation of repeating functions; used carefully and in obvious cases
Data-driven to insight-driven	Insight-driven when needs arise	Data-driven for cases with known needs; insight-driven for special cases, like decision support in solving complex problems
Latency—real time to right time	Ad hoc latency, as resources permit	Time tuned to the actual speed of business and required reaction times

Source: author

BI dimensions, presented in Chap. 4, find their reflection in maturity models (Table 8.1):

From the above attempt to relate BI dimensions and maturity, we can note that immature BI is reflected by limited approaches, while a more mature BI encompasses more dimensions and activities. Understandably, the expectations of BI owners land in the area of functional abundance and diversity, reflected in the ultimate levels or stages of maturity. The analysis of existing and well-known BIMM in Table 8.2 attempts to look at the ultimate levels of maturity for a group of the well-known BI maturity models, and detect possible patterns in different approaches to what kind of BI is considered mature.

All ultimate stages of BI maturity models, presented above, have the same mission—to evaluate the accumulated expertise in the area, assess current situation, estimate a trend and extrapolate it towards some ideal state. In some models, mature phase BI is everything to everyone on all dimensions. If evaluated along these guidelines, maturity represents a fact, or a point in time where the BI system reaches its goals driven by expectations. An interesting feature of most maturity models is the location of present time and situation in model steps, levels or phases—quite often the "now" is located at one phase from the last one. Thus, if a BIMM contains 6 levels, the first 4 levels reflect available experience, level 5 relates to current situation, and the ultimate level 6 is to express expectations towards an ideal state. There appear to be two main types of what is considered an ultimate level of maturity:

– Optimized—stable, regulated and efficient; many issues are formalized and managed;

Table 8.2 Features of well-known BI maturity models

Maturity model	Descriptions of the top maturity level	Type and features
Business intelligence maturity hierarchy (Deng, 2007)	Level 4—**wisdom**. It is the highest maturity level that every organization is striving to reach, enabling insightful applications of the knowledge available and changes in the business processes, and business productivity at this level should improve substantially because of timely and efficient decisions. In addition, we can note that IT role is less pronounced with every next maturity level	IT and infrastructure fade into the background
Gartner (Gartner's Business Intelligence and Performance Management Maturity Model, 2008)	The fifth stage is **pervasive**. Here the BI systems are integrated into business processes, and they present the required and reliable information to the users when changes happen. Employees of different levels can access information, perform data analysis, discover previously unknown issues and make decisions, including real-time decisions using real-time information. In other words, the BI process becomes more democratic and closer to the primary business processes. The implemented metrics are not only uniform throughout the organization, but are open to the clients and suppliers, thus reinforcing business relations. The main goal of the fifth stage is to maintain leadership even when major events occur, such as mergers, acquisitions or changes in management	BI integrated into business processes. Uniform metrics implemented Agile, adaptive and democratic BI nature Real-time decisions Reliable and viable infrastructure
TDWI model (Eckerson, 2007)	Level 5—**Sage**: the main features of this level are: *distributed development*, *data services*, and *extended enterprise*. The most typical usage of the BI system is creation of customized reports, KPIs, and other information services. The central unit is responsible for	Central information management; distributed development and customized services Regarding flexibility, it remains unclear which part of the system is most sensitive to changes when they occur,

(continued)

Table 8.2 (continued)

Maturity model	Descriptions of the top maturity level	Type and features
	management of the enterprise data warehouse as a repository for all enterprise information, while the development of customized solutions is distributed. For faster development of solutions, service oriented architecture (SOA) is used. A number of users is dramatically increased. Business and IT are aligned and cooperative. BI provides high-value services and competitive advantage (Hribar Rajterič, 2010)	and therefore should be most flexible
EBIM—Enteprise Business Intelligence Maturity (Tan et al., 2011)	Level 5—**Optimizing**: • Single view of truth—sources of information quality problems have been recognized and impact of poor quality calculated; • Enterprise data convergence—a single hub propagates data changes to all the application systems that need master data; • Warehouse analytical services—DW fades into the background as a data service and a part of BI infrastructure that acts as basis for a variety of analytical services; • Analytical competitor—continuous analytics review and enhancement	High centralization and accent on single view of truth. Optimizing approach
AMR Research (Shaaban, Helmy, Khedr, & Nasr, 2011):	Level 4—**orchestrated**: Intelligence strategy definition is done in a top-down approach. The goal is a common, agreed and rational view of the organization. Organizations detect important changes, and accordingly adjust the model and implementation to the slightest changes in the dynamic markets. Business runs based on measurable success factors. Cultural aspects are considered important for BI	A top-down approach to achieve an unifying optimized setup of intelligence processes

(continued)

Table 8.2 (continued)

Maturity model	Descriptions of the top maturity level	Type and features
	along with the technical aspect (Hribar Rajterič, 2010)	
HP Maturity model (The HP Business..., 2007)	Stage 5: • The organization benefits from an agile information environment, in which information no longer prohibits strategic agility but promotes and enables **strategic agility**; • Analytics are seen as a key differentiator for the organization, not just as a value-adding activity; • Users at all levels of the organization have access to insights that help them to work more effectively; • Integrated information is available seamlessly without regard to data source or integration technology used; • Unstructured and structured data are fully integrated; Advanced BI is fully embedded within processes, systems and workflow. • Information delivery efforts are characterized by agility, and new analytics are easily developed for new roles; • Once an organization reaches Stage 5 it must continue to evolve its BI capabilities by understanding and leveraging new technologies; • The exceptional organization must embrace an ongoing commitment to BI innovation and ensure that the information needs of users at all levels of the organization are met	Agile approach that accentuates flexibility
LOBI (Cates et al., 2005)	Level 6—**enabled intuition:** "an understanding of fundamental principles embodied within the nature of a complex real world challenge. It augments the intuitive reasoning power of the human mind"	Accentuates optimization and efficiency

(continued)

Table 8.2 (continued)

Maturity model	Descriptions of the top maturity level	Type and features
Business Analytics Capability Maturity Model (BACMM; Cosic, Shanks, & Maynard, 2012)	Level 4—**optimized**: the capability is at its maximum	Accentuates optimization
Business Information Maturity Model (Williams & Thomann, 2003)	Level 3 The organization recognizes the fact that decision processes initially are not optimal, and seeks to replace them with new decision processes, which **optimize** the usage of information throughout the whole organization	Accentuates optimization of decision processes
IBM Cognos BI Excellence Strategy Framework (Boyer, Frank, Green, Harris, & Van De Vanter, 2010)	Level 4: • Connecting strategy and execution • Formal organizational approach in place • IT, finance and business working together • Technology, people, and process standards for BI and performance management are in place	Leans toward optimization

Source: author

- Agile—adaptive, flexible, able to face changes; migrates from IT function to business function.

The most common features for optimization-type models are automation of high-level information activities, including automatic optimization; real-time reaction to changes in the environment; universal centralized standards. BI maturity here is mostly seen as an information system/function maturity. However, features required in a dynamic business environment, but omitted in many models, are flexibility, agility, ability to adapt while maintaining core competencies. Such features are accentuated in agility-type models that orientate towards organization maturity to leverage BI potential.

With each next stage, the journey of BI adoption brings in better understanding, structure, standards and manageability. On one hand, this managed infrastructure assures transparency and stability in informing processes. There are features in the ultimate stages for both types for models that are recognized necessary or *positive* for effective BI: reliable infrastructure, uniform set of company-wide standards and metrics. However, the excessive standardization and over-regulation eventually may develop BI features that are recognized as *negative* or doubtful: excessive automation of intelligence functions, rigid "optimized" structures, high level of inflexibility. Such features lead to diminished agility and inability to change.

There are ample reasons why these features are considered negative for BI: fluid and ever-changing subject area that is hard to put under a single integrated system, and requires flexible, adaptive and agile approaches; the set of company-wide standards should be carefully built to avoid rigidity; data integration from external and often unstable sources; review of information needs and change management. BIMMs that stress optimization and efficiency omit the fact that, for an optimized, and therefore strained, activity model, any unpredicted blow can bring grave consequences for the whole organization. An optimized model is tense and has little slack and flexibility, eventually becoming a source of severe risk—failure to react to any significant change could seriously undermine intelligence activities.

Based on current experience of BI use, a more balanced view on the concept of BI maturity is required to counter the often-inflated expectations and to survive inevitable changes in the future. In this book, approach to maturity is based on agility over optimization, regarding agility as a vital feature of contemporary business and its information environment, and largely connects with dynamic capabilities approach. BI maturing creates competencies that should be preserved, upgraded, and transformed to withstand changes in environment.

8.3 Redefinition of Business Intelligence Maturity

For BI maturity models to become more realistic, especially regarding their final phases, more realistic criteria are required. If the goal is to optimize intelligence and decision making, a problem of choosing optimization criteria emerges: are the criteria just internal, if BI addresses internal processes, or external criteria are used as well? The use of external criteria renders optimization impossible because of changing context. For this reason, the users of BI have to make up their mind whether they are after optimal decisions or just good enough decisions, whose search requires less time and effort. Here we can quote Snowden and Boone: "In a chaotic context, searching for right answers would be pointless" (Snowden & Boone, 2007). Information technologies can adapt to changing situations and be programmed to work flawlessly in a certain area, but in unknown environment that programmers have not encountered before, IT won't be able to function appropriately.

One more issue with BI maturity is that in mature phases of most maturity models BI is expected to operate smoothly in close connection with MIS/ERP systems. In our opinion, hard coupling of MIS and BI reduces agility of BI and informing in general. Although MIS/ERP and BI should work in tandem, the drive to make BI run as smooth as ERP systems are supposed to run is most likely to encounter problems—the mission of BI systems is much less structured, more sensitive to changing conditions and gives a larger role to human factors.

We can make an assumption that BI as a fluid and ever-changing phenomenon will never be mature in a sense of stability and routine; then the very idea of BI maturity may be questioned. BI maturity seems to be a questionable concept that essentially deals more with BI adoption stages, and less with its life cycle once it had been adopted. The other option to define a concept of BI maturity might be a review of a separate BI trend for its maturity life cycle: some trends emerge, gain adoption and experience, become established and routinized, or fade away and are replaced by other emerging trends that are at the beginning of their maturity cycle.

For a mature technology or system, one hardly notices its presence—its existence is taken for granted. As an example, data base technology is mature, and has matured long ago. Compared to data base technology, which is more of an infrastructure, BI is at the forefront where important things happen, and for large part is expected to deliver products of concentrated value. The stable part of BI might become a part of BI infrastructure—repetitive and commonplace procedures which will remain unchanged as long as they use stable data sources and serve stable information needs. However, the changing context of activities and the resulting variety of information environment undermine this stability.

An alternative way to define the concept of BI maturity would be the introduction of an overarching term of "BI culture" that has to be developed towards the final maturity phases. In various BI maturity models, the common feature of this culture is wide acceptance of BI throughout the organization, and incorporating BI functions into everyday activity. Most BIMM lean towards ideal mechanistic organization, automation of intelligence processes and often with automation and optimization of key business activities in real time. In author's opinion, a much more realistic case would be the development of BI culture as a sustainable binding set of agreed norms and competencies to leverage the ever-changing intelligence potential.

8.4 Business Intelligence Agility

8.4.1 The Notion of Agility

The first part of the current chapter discussed the notion of mature BI, and presented arguments why the accumulation of the experience of BI use should develop features like flexibility, adaptability and ability to maintain key competencies under changing context. The term of BI agility is seemingly appropriate here, as it covers all named features. BI maturity and agility are often mentioned in close context, as it has been shown in the first part of this chapter. While BI maturity concept states that important known questions about creating value have been answered, the agility concept accentuates readiness for important unknown questions. The known BI maturity models focus on BI acceptance dynamics, when BI agility focuses on the dynamics of established BI in a turbulent environment. To author's knowledge, there are no established, or at least suggested, models of agility that would stage the process of becoming agile in phases similarly to BI maturity models.

The agility issues in the area of advanced informing have been raised in the early days of decision support systems by Keen (1980), who discussed DSS adaptation to unanticipated usage patterns. According to Evelson (2011), while earlier-generation BI technologies have matured into industrial-strength solutions—function-rich, scalable, and robust, they have largely failed to address one simple, pragmatic business reality: the need for flexibility and agility.

The term of agility is actively discussed in business context for the last couple of decades, although agility has always been an essential feature for businesses operating in a dynamic environment. The dominating notion of agility is the ability to swiftly leverage business opportunities, but other interpretations exist—e.g., to survive emerging threats and handle inevitable changes with possibly minimal losses (Amos, 1998; Dove, 2001; Goldman, Nagel, & Preiss, 1995; Prahalad, 2009; Sull, 2009). The ability to respond is recognized as essential to survival (Kuilboer, Ashrafi, & Lee, 2016). To preserve and improve performance, organizations need to be able to adjust their processes, products and behaviors along with the constantly alternating business environment to remain competitive and relevant on the market (Smith, 2002). Agility is not necessarily related to direct creation of value, but it is expected to create necessary flexibility and resilience for future opportunities and threats, and can be seen as a form of insurance against future changes in the environment.

Several examples of situations requiring agility:

- A key supplier is suddenly out of business;
- An ERP system goes down for a day;
- A business opportunity requires to drastically rearrange resources;
- An innovation by competitor renders your product obsolete.

Many sources see agility as a feature supporting awareness, fast reaction, flexibility, nimbleness. In addition to the above, there are traits of resilience and survivability that are closely tied with agility, and come into use during crisis and downfalls. Several aspects of agility in general are discussed below.

Agility as Agile Development Process

There are sources that discuss BI agility in terms of sprint-like work processes, and project this meaning of agile into BI development phase (Linders, 2013; Tanane, Laval, & Cheutet, 2018). This is not the angle pursued in this book; however, even in this case, according to Linders (2013), organizations that are "fit" to adopt agile generally need to be empirical, collaborative, self-organizing, and cross-functional. These are the traits of organization culture that relate very much to agile BI (Skyrius & Valentukevičė, 2020). This type of agility accentuates a specific setup of work processes.

Agility as Fast Reaction

Numerous sources relate agility, including BI agility, as an ability to move and react fast. In a study by Aberdeen group (White, 2011), BI agility is understood as rapid reaction, nearing real-time BI. Knabke and Olbrich (2013) analyzed published research for definitions of agility in BI context, and found out that many definitions center around agility as nimbleness out of fast responses. Sengupta and Masini (2008) discuss IT agility as the ability of a firm to adapt its IT capabilities to market changes, and suggest two types of IT agility: range agility—ability to grow or shrink the external (products and services) or internal (capacity, structure) repertoire; and time agility—the speed of response. The authors indicate that range agility is more relevant for stable markets, and time agility—for more dynamic ones. This type of agility stresses comprehensive informing on current issues, fast detection and reaction.

Agility as Resilience

Another often-mentioned aspect of agility is resilience, commonly understood as an ability to withstand the turbulent cycles and maintain key competencies by recombining assets. Sengupta and Masini (2008) stressed range agility as a capacity to change the product range or internal processes to survive aggressive changes in the environment. Gulati (2010) researched how resilient companies prosper both in good and bad times, cutting through internal barriers that impede action, and building bridges between warring divisions. Based on more than a decade of research in a variety of industries, and case studies from companies including Cisco Systems, La Farge, Starbucks, Best Buy, and Jones Lang LaSalle, Gulati explored the five levers of resilience:

- Coordination: Connect, eradicate, or restructure silos to enable swift responses.
- Cooperation: Foster a culture that aligns all employees around the shared goals of customer solutions.
- Clout: Redistribute power to "bridge builders" and customer champions.
- Capability: Develop employees' skills at tackling changing customer needs.
- Connection: Blend partners' offerings with yours to provide unique customer solutions.

While resilience levers like capability and connection belong to common, if essential, competencies, it is the trio of coordination, cooperation and clout that aims at profound cultural change and removal of local fiefdoms. A dynamic capabilities approach, proposed by David Teece and others (Eisenhardt & Martin, 2000; Teece, 2007; Teece, Peteraf, & Leih, 2016; Teece, Pisano, & Shuen, 1997) followed a related path, based on reconfiguration of competences in changing environments. Agility is seen as a competence meant not only to outperform competitors, but to survive in turbulent and often aggressive conditions.

This type of agility stresses organizational vitality, supported by specific assets and competencies that hold their importance over time.

Non-agile Qualities

The above features of agile activities point towards some desired state—faster, more flexible, resilient, sustainable etc. Sometimes it helps to better define a term by looking at the opposite pole of the discussed dimension. In the case of agility it might provide a better understanding of migration towards complacency and self-satisfying representation of reality. Here we can take a closer look at what is commonly deemed as non-agile. Some sources (Oetringer, 2018) point at increasing complexity of activities that exceeds a tipping point, and the strategies, standards, processes and best practices start lagging behind reality. On the other hand, success breeds complacency and reluctance to change, and lack of success increases the development and use of dynamic capabilities (Zahra, Sapienza, & Davidson, 2006). Mayer and Schaper (2010) discussed inflexibility between architectures of business and IT, stressing the rigidity of established information systems that limits the capture, organization, and accessibility of data. Such systems are ill-equippped to handle external updates like regulatory changes or reporting standards. As an alternative, Excel-based solutions emerge as a parallel informing environment.

Gill (2008), discussing relation between structural complexity and effective informing, presented cases of Mrs. Fields Cookies and Batterymarch, where the initial successful projects for effective use of IS eventually morphed into inadaptability to adapt to environment changes. In both cases, the highly specialized systems optimized for a narrow functionality appeared to be highly vulnerable to changes in the environment. The benefits gained from exclusive focus came in at a price of eliminating individuals whose role included environmental sensing and detecting important changes in the environment.

Traditional efficiency-oriented business models, when evaluated against agility issues, are non-agile in most aspects (Parr Rud, 2009). Such models are based on assumption that all individual or system behavior is predictable and controllable; the accent is on efficiency and optimization; and in unpredicted situation root cause search is performed. In a highly volatile environment, such precisely balanced (and therefore strained) models carry the risk of inflexibility and expensive changes.

Although the topic of this subchapter is BI agility, it is not a goal by itself—the ultimate value is the agility of an organization. Therefore, three levels of agility—organizational agility, informational agility and BI agility will be discussed further in this chapter together with their interrelations.

8.4.2 Organizational Agility, Information Agility and BI Agility

Organizational agility can be defined as a set of processes that allows an organization to sense changes in the internal and external environment, respond efficiently and effectively in a timely and cost-effective manner, and learn from the experience to improve the competencies of the organization (Seo & LaPaz, 2008). In turbulent environments that have become a norm for business activities, much of organizational agility rests on the competencies to regroup and recombine assets for appropriate action. Such competencies largely relate to dynamic capabilities approach (DCA), proposed by Teece (2007) and others. This regrouping would be impossible without another element of agility foundation—awareness through effective informing activities to produce and use quality information. There are multiple cases with above elements having positive or adverse effects on organization agility and, in some cases, on its survival.

Changes in firms are spurred by failure in three steps: recognition of the failure, interpretation of results as failure, and adjustment of capabilities (Zahra, Sapienza, & Davidson, 2006). Changes in external circumstances force invention of solutions in order to survive; such solutions may be not in the usual set of routines. So firms engage in experimentation, leaning-by-doing, trial-and-error learning, and improvisation. Selection of the path depends on firm age and the development level of their routines.

Comparing traditional business models with agile models, Parr Rud (2009) noted that rather than implementing current change, the agile model (although the author has used the term "emergent" instead of "agile") is always adapting to stay in balance. The continuous turbulent business climate demands a view that supports adaptability and resilience. In addition to that, the intelligence of the organization does not just reside with leadership but is assumed to be cross-functional and distributed across a wide variety of people and systems, as an adaptable organization possesses a culture that fosters self-organization.

Keller, Ollig, and Fridgen (2019) discussed the relation between organizational agility (flexible and resilient) and reliability (stable), and brought in several specific terms related to organizational agility and its relation to IT:

- **Mindfulness**—dynamic awareness that concentrates attention to important attributes like failure or resilience, perception of complete context and creation of new options; or
- **Meta-routines** that enable changes, performance of non-routine tasks, and changes in other routines that do not cover unexpected things.

Regarding the tradeoff between agility and stability, organizational agility is expected to contribute to stability by maintaining desired level of operation regardless of the disruptions. However, responding creatively is often not possible with traditional routine procedures. Standardization slows down reaction to opportunities and threats. On the other hand, a permission for local units to act freely and flexibly may significantly undermine the global reliability. Decoupling of systems can enable agility in short-term perspective, but undermine long-term reliability and agility by increasing "technical debt" (difference between current and desired state of IT or IS). To balance this tradeoff, a portfolio-like approach to balance technologies and their configurations to achieve greater agility has been suggested by Park, El-Sawy, & Fiss (2017).

Information agility. Organizational agility is largely ensured by information agility or informing agility (Gill, 2008). Information systems were always expected to enhance agility of an organization. As early as 1997, Sambamurthy and Zmud defined IT capability as "a firm's ability to acquire, deploy, combine, and reconfigure IT resources in support and enhancement", thus strongly supporting the informing agility competence. The ultimate goal can be named the agile information organization (DeSouza, 2007).

The importance of information systems and informing activities in developing organizational agility brings in considerable challenges. Lu and Ramamurthy (2011) state that firms invest in information technology to pursue fast initiatives in response to constant changes in environment, and indicate a lack of understanding of contradictions between IT and agility in the general context of IT use. Making organizations agile can be challenging and costly due to different business models, organizational structures, IT systems, and investments to support the IT-organizational agility relationship (Bani Hani, Deniz, & Carlsson, 2017). Timo Elliott, a SAP evangelist (Elliott, 2014), noted in his blog: "One of the ironies of analytics today is that while the technology is more powerful and easier to use than ever, the **fast-changing** organizational and technical landscapes have lead to increased discontent and confusion for IT and business users alike". Although IT resources are critical for quick dissemination of information and decision speed, IT infrastructures themselves are often rigid and inertial (Sengupta & Masini, 2008). The contradiction between the desire to be agile with pressure to streamline IT and its spending brings up some difficult decisions (Tallon, Queiroz, Coltman, & Sharma, 2019).

Seo and LaPaz (2008) have investigated the risks associated with relation between information systems and organizational agility, and indicated a set of IS features having adverse effect on this relation; some of the most agility-related features are:

- Lack of integration between systems—a result of incremental IS development.
- Unstandardized raw data (no data integration, company-wide standards or SST).
- Scope of processing: inability to cover or process all data—important facts or trends can be overlooked; the opposite—if the data is over-generalized or "overfit" to standards, important context may be lost.
- Rigid IS architecture may lead to missing important signals.
- IS inflexibility: stable IS can quickly become rigid (the eternal conflict between efficiency and flexibility).
- Greater propensity for error: an error, incorrect or failed transaction can be amplified further the chain of linked information processes, as the case of sporting goods company Nike shows (Koch, 2004).

Apparently, the significant share of problems in seeking agility through informing activities lie in the inertia and rigidity of foundation information systems, be they of the type of legacy systems, or "systems zoo" that came into being over multiple system procurements, or effective yet rigid and limited systems. Information infrastructure contributes to agility by being stable and having as few limitations as possible. As its life cycle is expected to outlast several waves of BI applications, a reasonable set of conventions and standards would contribute to both agility and stability.

BI agility. In discussing agility issues, more and more sources relate advanced informing functions and business intelligence in particular as a key factor in building up agility. The role of business intelligence is of prime importance here because it provides the required insights on own activities and their environment, and serves as a potential lever for organizational agility (Baars & Zimmer, 2013). Such insights reach beyond the role of traditional information systems, providing essential functions of monitoring, early warning and decision support. So the agility of BI systems and activities may be defined as the capacity to maintain and reconfigure the portfolio of important informing competencies that provides maximum resiliency or scalability for future trends or opportunities. As BI has to stay in tune with dynamic environments, Scheps (2008) ironically noted that BI is like an infant: it can't operate unattended for very long without starting to smell funny.

BI has to provide support in two ways: stable and reliable reporting; and agile future-oriented analytics (Knabke & Olbrich, 2016). Because BI systems not only are supposed to respond to changes, but to foresee them as well, the need for agility inevitably pushes them towards some degree of de-centrality. The BI development project in fact becomes a permanent process, and the BI state becomes a process as well. According to Zimmer, Baars, and Kemper (2012), there needs to be a balance between discipline (rules, standards) and agility; once more, it can be added that in search for greater agility, the coverage of rules and standards should be rather basic and clear.

The growing interest for the role of BI agility may be explained, at least in part, by growing interest for survival strategies, and among sources of research on BI, agility is mentioned as one of the key BI features. As Zimmer et al. (2012) state, agility is central to dynamic environments. The competitive survival strategies, based on dynamic capabilities, start to dominate over competitive advantage strategies that are based on resource-based view, and this factor boosts the importance of agility.

While BI is seen by many as the instrument for maintaining awareness, and therefore absolutely essential for organizational agility, the strictness of the business rules and regulations supported by the BI directly affects the flexibility of BI (Işık, Jones, & Sidorova, 2013). If strict rules and regulations are embedded in the applications, BI flexibility will diminish and impair the ability to deal with exceptions and urgencies.

Describing the case of agile BI at Disneyland Paris, Iafrate (2013) makes an important point: in intelligence applications, patience is required to see the potential value, but it is also important to get something useful running quickly, to get the culture changes started and provide continuous value. In Disneyland Paris, the

Table 8.3 Environment changes inducing changes in environment monitoring procedures (unpublished research data)

Environment changes	No. of cases
New channels of receiving information	139
New conditions of one's activities	138
New approaches to insight	75
Other, e.g.: – Changes in supplier IS; – Changes in economic and political situation; – Research before introducing a new special product.	18

models are constantly updated and become more accurate as the day or week progresses and the impact of unusual patterns can quickly be understood and adjustments made, adding to the flexibility and agility of BI system. This approach also requires intense communication and sharing of information and insights.

In an earlier unpublished research survey by the author, performed in 2009, a question about what environment changes induce changes in environment monitoring procedures yielded the results presented in Table 8.3 out of 204 participants (a convenience sample of small and medium business managers).

The responses give a hint on drivers of intelligence agility, or ability to adapt to changes in environment. The data tell that external changes (new channels, new conditions) force to change monitoring modes because of their inevitability, while new approaches to insight (changing the ways users interpret data and information) do not affect the monitoring modes much. This might be due to the fact that users tend to use the same raw data and information but with different approaches that come from the human side of monitoring (sense making), while new channels can be attributed to IT support and require new technical skills.

8.5   Dynamic Capabilities Approach

As stated earlier in this chapter, the notion of organizational agility is closely related to the dynamic capabilities approach (DCA). Teece's concept of dynamic capabilities (Teece et al., 1997) essentially says that what matters for business is corporate agility: the capacity to sense and shape opportunities and threats, seize opportunities, and maintain competitiveness through enhancing, combining, protecting, and, when necessary, reconfiguring the business enterprise's intangible and tangible assets. Dynamic capabilities are defined as the abilities to reconfigure a firm's resources and routines in the manner envisioned and deemed appropriate by its principal decision makers (Zahra, Sapienza, & Davidson, 2006). According to Knabke and Olbrich (2016), the dynamic aspect refers to "the ability to renew competences according to changing environments", reflecting reactive position to be able to adapt to changes. However, agility, in our understanding, encompasses the proactive side as well, and includes the built-in strengths that provide for easier adaptability in future changes.

DCA essentially ties together BI and competitive advantage, or competitive survival strategies. The BI agility, based on flexible approaches, seems to be a definitive factor in enhancing dynamic capabilities. We have to note that DCA orientates BI towards a needs-based approach: data-driven approach sets its focus off-target by concentrating on possible data derivatives without clear business value; with needs-based approach, insights and sense are the focal point. According to (Teece, 2007), sustainable enterprise has the capacity to: (1) sense and shape opportunities and threats, using a mix of external and internal intelligence, (2) seize opportunities, and (3) maintain competitiveness through enhancing, combining, protecting, and, when necessary, reconfiguring the business enterprise's intangible and tangible assets and using mostly internal intelligence. Following this approach, agile BI function, as one of the key intangible assets, should possess the required agility to reconfigure itself when necessary, with minimal costs or losses.

8.6 Factors Supporting and Restricting Agility

In order to gain a better understanding of agile environment it is important to be aware of factors not only supporting agility but also the ones restricting it. The summary of factors supporting or restricting agility, detected in literature analysis, is presented in Table 8.4.

From the data in the above table, we can see that there is a significant set of information-related factors influencing organizational agility, both at the levels of general informing functions (information agility) and BI functions (BI agility). Some of the factors are recognized as technical or 'hard' qualities (e.g., solution architecture), and the others are human related or 'soft' ones (e.g., culture). In the terms of organizational preparedness for changes, BI activities are expected to support it by providing the required monitoring and sensing of the environment, so it is easy to assume that agile BI will maintain its key competencies for longer time with required levels of performance. This point is supported by Van Beek (2014), who stated that one of BI agility traits is the BI reuse. We may also infer that organizations with flatter structures and less rigid hierarchies possess a type of culture and climate that is more supportive for agility.

8.7 BI Agility Management Issues

BI agility, as well as agility in general, is a "soft" concept, dealing with often unpredictable future events, or forward-looking (Tallon et al., 2019), and therefore hard to manage directly. Among possible indirect agility management approaches

Table 8.4 Factors supporting or restricting agility (Skyrius & Valentukevičė, 2020)

	Supporting factors	Restricting factors
Organizational agility	• Agile organizational culture—flexible, creative, error-tolerant (Bateman & Snell, 2012; Linders, 2013; Tallon et al., 2019) • Timely detection of important events and cues (Lee, Sambamurthy, Lim, & Kwoh, 2015; Overby, Bharadwaj, & Sambamurthy, 2006; Roberts & Grover, 2012; Seo & LaPaz, 2008; Appelbaum, Calla, Desautels, & Hasan, 2017; Tallon et al., 2019; Chen & Siau, 2012) • Rapid learning (Appelbaum et al., 2017; Oetringer, 2018) • Fast decisions (Appelbaum et al., 2017; Queiroz, Tallon, Coltman, & Sharma, 2018) • The ability to renew competences according to changing environments (Knabke & Olbrich, 2016) • Mature BI capabilities (Watson & Wixom, 2007) • Identification of core competencies (Bateman & Snell, 2012)	• Rigid, conservative organizational culture (Bateman & Snell, 2012) • Industrial organization (Verheyen, 2019) • Information overload for decision makers (Seo & LaPaz, 2008) • Lack of integration between information sources and perception systems (Seo & LaPaz, 2008) • Variety of IT systems (Bani Hani et al., 2017) • Inflexible IT systems (Queiroz et al., 2018; Van Oosterhout, Waarts, & Van Hillegersberg, 2006) • Large organization size (Daft, 2001)
Informing or information agility	• Agile informing culture (Choo, 2013) • Emergent coordination (Liang, Wang, Xue, & Ge, 2017) • Removal of departmental information silos (Imhoff & Pettit, 2004) • Flexible IT infrastructure (Appelbaum et al., 2017) • Clear, rational and well-defined basic informing processes—reporting, KPI (Lee et al., 2015; Oetringer, 2018) • Enterprise-wide information sharing (Liang et al., 2017) • Capability to reconfigure IT resources (Sambamurthy & Zmud, 1997) • Dynamic nature of BI (Mayer & Schaper, 2010; Zimmer et al., 2012)	• Messy, chaotic IS strategy (Lu & Ramamurthy, 2011; Seo & LaPaz, 2008) • Bad data quality (Seo & LaPaz, 2008) • Departmental information silos (Imhoff & Pettit, 2004; Tallon et al., 2019)
BI agility	• Agile BI culture (Bateman & Snell, 2012) • Network structure (Appelbaum et al., 2017) • Information and insight sharing (Skyrius, Nemitko, & Taločka, 2018) • Lessons and experience (Oetringer, 2018) • Community of experts/analysts (Appelbaum et al., 2017)	• Rigid standards (Seo & LaPaz, 2008; Tallon et al., 2019) • Information overload (Tallon et al., 2019)

that may emerge from the previous discussion, two more obvious approaches emerge: through individual agility of BI assets; and through nurturing a supportive type of culture that encompasses BI assets.

Agility management by BI assets. It should encompass how well the current assets will hold over changes; how easy it would be to adjust, replace or extend them during changes; and how easy it would be to regroup or rearrange them. Baars and Zimmer (2013) propose a set of indicators covering sources of BI agility: content agility—time and efforts to make changes in information sources; functional agility—time and efforts to make changes in functionality; and scale agility—time and effort to make changes in BI capacity and throughput. This seemingly simple approach is to become rather complicated when getting into more detail regarding possible sources of changes in BI assets:

- Information sources and changes in their availability and access; the multiplicity of publicly available sources and criteria for rational selection of sources and available information in general; discovery of new sources; criteria for a portfolio approach for source selection;
- Procedures of handling information—processing and analytics, matching and integrating, distribution and evaluation; advances in information handling that may make old procedures partly or entirely obsolete; evaluation of IT innovations; changes focusing more on technical procedures or human procedures—approvals, granting access rights etc.;
- Utilization of new informing environments or other radical innovations on approach like social networks in public or private information space;
- Capacity and its measures in user terms rather than in technical terms; the level of satisfaction of users' needs for information and insights should be measured, although the possible units for such measure—quality decisions or valuable insights—are hard to define;
- People, their competencies and expertise—alongside with individual competencies, teamwork and joint effort are of prime importance; Seo and LaPaz (2008) state that hiring agile people might be not enough without proper organizational structure and culture.

Not all of BI assets should be flexible and ready to adapt. There are certain stable BI assets that hold their value over time and alleviate change. Several examples of these assets may be: important and stable information sources with delivery channels and methods; support of archives and own information collections; organizational structures for sharing and cooperation; self-service BI and analytical skills, to name a few. Their value justification requires time and changing environment.

In many sources of research on agility, used in this chapter, the traits of agility lead to cultural issues. In an extensive literature review on organizational agility by Tallon et al. (2019), the antecedents or enablers of agility can be subdivided into four general categories: technological, behavioral, organizational/structural, and environmental. The culture, which falls under behavior category, and partly under organizational/structural category, is one of the essential drivers for change and is core of every organization with linkages to leadership, behaviors, org structure, processes, policies.

Agility management through culture. As shown several times in this chapter, together with information, culture plays important role in enabling organizational agility. One of the most accurate definitions of organizational culture comes from Schein (2016) who describes it as: the accumulated shared learning of the group as it solves its problems of external adaptation and internal integration. This accumulated learning is a pattern or system of beliefs, values, and behavioral norms that come to be taken for granted as basic assumptions and eventually drop out of awareness.

The types of cultures that enable greater agility are being rooted in people-focused attributes such as autonomy, empowerment, collaboration. An environment, underpinned by such attributes and values, enables organizations, teams and individuals to be more adaptive, flexible, innovative and resilient when dealing with complexity, uncertainty and change (Gogate, 2017).

In various sources authors often stress that agile is more culture than process (Denning, 2010). Agile culture is often associated with the following characteristics (Kulak & Li, 2017; Hesselberg, 2019; Spayd, 2010; Appelo, 2010):

- Shared sense of purpose;
- Servant leadership;
- Self-organizing and empowered teams;
- Collaboration and communication;
- Rapid iteration and experimentation;
- Continuous improvement and learning;
- Learning from failure;
- Willingness to share knowledge;
- Trust and transparency.

At the core of agile culture is acceptance that this kind of culture should always be learning, developing and evolving so an end goal as a stable point in time should never be reached.

The organizational culture corresponds to the values and beliefs of the company's people, guides their behavior, and influences how the company operates, including the ways of work with information and informing processes—also identified by researchers as information culture. The information culture is a part of organizational culture that is concerned with the assumptions, values, and norms that people have about creating, sharing, using information (Choo, 2013). We may assume that part of the information culture is BI culture which supports the complex part of information needs scale intended to satisfy the needs for insights and decision support that arise from important and possibly costly issues (Skyrius et al., 2018).

If organization is characterized as agile, this culture will reflect in the way company works with information and uses BI applications to gain better insights. Popovic, Hackney, Coelho, and Jaklic (2012) state that an analytical decision-making culture appears to be a critical factor in ensuring BIS success. While BI capabilities are associated with higher organizational agility, it is equally important that managers promote a culture of calculated risk taking in order that employees have a way to test their ideas without fear of failure or retribution from superiors.

Summing up, BI agility as a capacity to change and adapt while maintaining key competencies, together with information agility in general, is a key component in organizational agility both for competing and survival. A planned development of an agile BI system does not seem likely—it has to evolve and survive in a sustainable culture using assets that are deemed core assets at the time, and higher level competencies to reconfigure itself. A specific type of culture—BI culture should be nurtured around informing activities and BI activities in particular. The issues of BI culture are going to be discussed in more detail in the next chapter of this book.

References

Amos, J. (1998). *Transformation to agility: Manufacturing in the marketplace of unanticipated change*. New York, NY: Garland.

Appelbaum, S., Calla, R., Desautels, D., & Hasan, L. (2017). The challenges of organizational agility (part 1). *Industrial and Commercial Training, 49*(1), 6–14.

Appelo, J. (2010). *Management 3.0: Leading agile developers, developing agile leaders*. Boston: Pearson Education.

Baars, H., & Zimmer, M. (2013). A classification for business intelligence agility indicators. *ECIS 2013 Completed Research*. Paper 163.

Bani Hani, I., Deniz, S., & Carlsson, S. (2017). Enabling organizational agility through self-service business intelligence: The case of a digital marketplace. *The Pacific Asia Conference on Information Systems (PACIS)*, p. 148.

Bateman, T., & Snell, S. (2012). *Management: Leading & collaborating in the competitive world* (10th ed.). New York, NY: McGraw-Hill Education.

Boyer, J., Frank, B., Green, B., Harris, T., & Van De Vanter, K. (2010). *Business intelligence strategy. A practical guide for achieving BI excellence*. San Jose, CA: IBM Corporation.

Cates, J. E., Gill, S. S., & Zeituny, N. (2005). The ladder of business intelligence (LOBI): A framework for enterprise IT planning and architecture. *International Journal of Business Information Systems, 1*(1/2), 220–238.

Chen, X., & Siau, K. (2012). Effect of business intelligence and IT infrastructure flexibility on organizational agility. *Proceedings of the 33rd International Conference on Information Systems (ICIS)*, Orlando, pp. 1–19.

Choo, C. W. (2013). Information culture and organizational effectiveness. *International Journal of Information Management, 33*(5), 775–779.

Cosic, R., Shanks, G., & Maynard, S. (2012). Towards a business analytics capability maturity model. *23rd Australasian Conference on Information Systems*, 3-5 Dec. 2012, Geelong, Australia.

Crawford, J. K. (2006). The project management maturity model. *Information Systems Management, 23*(4), 50–58.

Daft, R. L. (2001). *Organization theory and design* (7th ed.). Cincinnati, OH: South West College Publishing.

de Bruin, T., Freeze, R. D., Kulkarni, U., & Rosemann, M. (2005). Understanding the main phases of developing a maturity assessment model. In D. Bunker, B. Campbell, & J. Underwood (Eds.), *Australasian Conference on Information Systems (ACIS)* (pp. 8–19). CD Rom: Australasian Chapter of the Association for Information Systems.

Deng, R. (2007). Business intelligence maturity hierarchy: A new perspective from knowledge management. *Information management*. Retrieved April 24, 2009, from http://www.information-management.com/infodirect/20070323/1079089-1.html.

Denning, S. (2010). *The leader's guide to radical management: Reinventing the workplace for the 21st century.* San Francisco, CA: Jossey-Bass.

DeSouza, K. C. (Ed.). (2007). *Agile information systems: Conceptualization, construction and management.* Oxford: Elsevier.

Dove, R. (2001). *Response ability: The language, structure, and culture of the agile enterprise.* New York, NY: Wiley & Sons.

Eckerson, W. (2007). Beyond the basics: Accelerating BI maturity. *The Data Warehousing Institute Research.* Retrieved October 30, 2019, from http://download.101com.com/pub/tdwi/Files/SAP_monograph_0407.pdf.

Eisenhardt, K., & Martin, J. (2000). Dynamic capabilities: What are they? *Strategic Management Journal, 21*, 1105–1121.

Elliott, T. (2014) *5 Top tips for agile analytics organizations.* Retrieved June 12, 2020, from https://timoelliott.com/blog/2014/09/5-top-tips-for-agile-analytics-organizations.html.

Evelson, B. (2011.) *Buyer's guide: How agility will shape the future of business intelligence.* Retrieved December 19, 2019, from https://www.computerweekly.com/feature/Buyers-Guide-How-agility-will-shape-the-future-of-business-intelligence.

Gartner's Business Intelligence and Performance Management Maturity Model. (2008). Retrieved November 8, 2019, from http://www.gartner.com/DisplayDocument?id=500007.

Gifford, C. (2011). *When worlds collide in manufacturing operations: ISA-95 best practices book 2.0.* Research Triangle Park, NC, USA: International Society of Automation.

Gill, T. G. (2008). Structural complexity and effective informing. *Informing Science, 11*, 253–279.

Gogate P. (2017) *Towards and agile culture—Agile Culture and Leadership. Agile Business Consortium. Interactive*: Retrieved from www.agilebusiness.com.

Goldman, S. L., Nagel, R. N., & Preiss, K. (1995). *Agile competitors and virtual organizations: Strategies for enriching the customer.* New York, NY: Van Nostrand Reinhold.

Gottschalk, P. (2002). Toward a model of growth stages for knowledge management technology in law firms. *Informing Science, 5*(2), 79–93.

Gulati, R. (2010). *Reorganize for resilience: Putting customers at the center of your business.* Boston, MA: Harvard Business Press.

Hallikainen, P., Merisalo-Rantanen, H., & Marjanovic, O. (2012). From home-made to strategy-enabling business intelligence: The Transformatorial Journey of a Retail Organization. *Proceedings of the ECIS-2012 European Conference on Information Systems.* p. 28.

Hesselberg, J. (2019). *Unlocking agility: An insider's guide to agile enterprise transformation.* Boston, MA: Addison-Wesley.

Hribar Rajterič, I. (2010). Overview of business intelligence maturity models. *Management, 15*(1), 47–67.

Humphrey, W. S. (1988). Characterizing the software process: A maturity framework. *IEEE Software, 5*(2), 73–79.

Iafrate, F. (2013). Use case: Business intelligence "new generation" for a "zero latency" organization (when decisional & operational BI are fully embedded). In P.-J. Benghozi, D. Krob, & F. Rowe (Eds.), *Digital enterprise design & management.* Berlin: Springer. Retrieved October 30, 2019, from http://link.springer.com/chapter/10.1007%2F978-3-642-37317-6_1.

Imhoff, C., & Pettit, R. (2004). *The critical shift to flexible business intelligence: What every marketer wants—and needs—from technology.* Retrieved October 30, 2019, from http://download.101com.com/pub/tdwi/files/Critical_Shift_to_Flex_BI_Imhoff.pdf.

Işık, Ö., Jones, M. C., & Sidorova, A. (2013). Business intelligence success: The roles of BI capabilities and decision environments. *Information & Management, 50*(2013), 13–23.

Keen, P. G. W. (1980). Adaptive design for decision support systems. *ACM SIGMIS Database: The DATABASE for Advances in Information Systems, 12*, 1–2.

Keller, R., Ollig, Ph., & Fridgen, G. (2019). Decoupling, information technology, and the tradeoff between organizational reliability and organizational agility. *Proceedings of the 27th European Conference on Information Systems (ECIS)*, Stockholm & Uppsala, Sweden, June 8–14, 2019.

Knabke, T., & Olbrich, S. (2013). Understanding information system agility—The example of business intelligence. *Proceedings of the 46th Hawaii International Conference on System Sciences*, pp. 3817-3826.

Knabke, T., & Olbrich, S. (2016). Capabilities to achieve business intelligence agility—Research model and tentative results. *Proceedings of Pacific Asia Conference on Information Systems (PACIS) 2016*, p. 35.

Koch, C. (2004). Nike Rebounds. *CIO Magazine*, June 15, 2004. Retrieved November 30, 2019, from https://www.cio.com/article/2439601/nike-rebounds%2D%2Dhow%2D%2Dand-why%2D%2Dnike-recovered-from-its-supply-chain-disaster.html.

Kuilboer, J.P., Ashrafi, N., & Lee, O.D. (2016). Business intelligence capabilities as facilitators to achieve organizational agility. *22nd Americas Conference on Information Systems (AMCIS) 2016: Surfing the IT innovation wave*, pp. 1–5.

Kulak, D., & Li, H. (2017). *The journey to enterprise agility: Systems thinking and organizational legacy*. Cham, Switzerland: Springer.

Lee, O. K., Sambamurthy, V., Lim, K. H., & Kwoh, K. W. (2015). How does IT ambidexterity impact organizational agility? *Information Systems Research, 26*(2), 398–417.

Liang, H., Wang, N., Xue, Y., & Ge, S. (2017). Unraveling the alignment paradox: How does business-IT alignment shape organizational agility? *Information Systems Research, 28*(4), 1–17.

Linders, B. (2013). *Adopting agile when your management style is mostly command and control*. Retrieved October 30, 2019, from www.infoq.com/news/2013/10/agile-command-and-control.

Lu, Y., & Ramamurthy, K. (2011). Understanding the link between information technology capability and organizational agility: An empirical examination. *MIS Quarterly, 35*, 931–954.

Mayer, J. H., & Schaper, M. (2010). Turning data into dollars—Supporting top management with next-generation executive information systems. *McKinsey on Business Technology, 18*, 12–17.

Nolan, R. (1973). Managing the computer resource: A stage hypothesis. *Communications of the ACM, 16*(4), 399–405.

Oetringer, E. (2018). Business Intelligence: Where are the lessons learned? Accessed on March 9, 2020, https://comdys.com/2018/02/22/bi/.

Overby, E., Bharadwaj, A., & Sambamurthy, V. (2006). Enterprise agility and the enabling role of information technology. *European Journal of Information Systems, 15*, 120–131.

Park, Y., El-Sawy, O., & Fiss, P. C. (2017). The role of business intelligence and communication technologies in organizational agility: A configurational approach. *Journal of the Association for Information Systems, 18*(9), 648–686.

Parr Rud, O. (2009). *Business intelligence success factors: Tools for aligning your business in the global economy*. Hoboken, NJ: John Wiley & Sons.

Popovic, A., Hackney, R., Coelho, P. S., & Jaklic, J. (2012). Towards BI systems success: Effects of maturity and culture on analytical decision making. *Decision Support Systems, 54*(1), 729–739.

Prahalad, C. K. (2009). In volatile times, agility rules. *Business Week, September, 21*, 80.

Queiroz, M., Tallon, P., Coltman, T., & Sharma, R. (2018). Corporate knows best (maybe): The impact of global versus local IT capabilities on business unit agility. *Proceedings of the 51st Hawaii International Conference on System Sciences (HICSS)*, Association for Information Systems (AIS), pp. 5212–5221.

Roberts, N., & Grover, V. (2012). Leveraging information technology infrastructure to facilitate a firm's customer agility and competitive activity: An empirical investigation. *Journal of Management Information Systems, 28*(4), 231–270.

Sambamurthy, V., & Zmud, R. W. (1997). *At the heart of success: Organization-wide management competences*. San Fransisco, CA: Jossey-Bass.

Scheps, S. (2008). *Business intelligence for dummies*. Indianapolis, IN: Wiley Publishing.

Sengupta, K., & Masini, A. (2008). IT agility: Striking the right balance. *Business Strategy Review, London Business School, Summer, 2008*, 42–47.

Seo, D., & LaPaz, A. I. (2008). Exploring the dark side of is in achieving organizational agility. *Communications of the ACM, 51*(11), 136–139.

Shaaban, E., Helmy, Y., Khedr, A., & Nasr, M. (2011). Business intelligence maturity models: Toward new integrated model. In: *The International Arab Conference on Information Technology* (ACIT 2011), Riyadh, Saudi Arabia, 11–14 December 2011.

Skyrius, R., Nemitko, S., & Taločka, G. (2018). The emerging role of business intelligence culture. *Information Research, 23*, 4. Retrieved October 22, 2019, from http://informationr.net/ir/23-4/infres234.html.

Skyrius, R., & Rėbždaitė, V. (2011). Business intelligence maturity and prospects. In *Proceedings of 2011-03-25 conference changes in economy and asset value*. Lithuania: Vilnius University.

Skyrius, R., & Valentukevičė, J. (2020). *Business Intelligence Agility, Informing Agility and Organizational Agility: Research Agenda. A manuscript accepted for publication at Informacijos Mokslai*. Lithuania: Vilnius University.

Smith, M. E. (2002). Success rates for different types of organizational change. *Performance Improvement, 41*(1), 26–35.

Snowden, D., & Boone, M. (2007). A leader's framework for decision making. *Harvard Business Review, 85*(11), 69–76.

Spayd, M. (2010). *Agile & culture: The results*. Retrieved March 2, 2020, from http://www.collectiveedgecoaching.com/2010/07/agile__culture/.

Sull, D. (2009). How to thrive in turbulent markets. *Harvard Business Review, 87*(2), 78–88.

Tallon, P. P., Queiroz, M., Coltman, T., & Sharma, R. (2019). Information technology and the search for organizational agility: A systematic review with future research possibilities. *Journal of Strategic Information Systems, 28*(2019), 218–237.

Tan, C.-S., Sim, Y.-W., & Yeoh, W. (2011). A Maturity Model of Enterprise Business Intelligence. *Communications of the IBIMA, 2011*, 417812.

Tanane F.-Z., Laval, J., & Cheutet, V. (2018). Towards assessment of information system agility. *10th IEEE International Conference on Software, Knowledge, Information Management and Applications* (SKIMA 2016), Dec 2016, Chengdu, China.

Teece, D. (2007). Explicating dynamic capabilities: The nature and microfoundations of (sustainable) enterprise performance. *Strategic Management Journal, 28*, 1319–1350.

Teece, D., Peteraf, M., & Leih, S. (2016). Dynamic capabilities and organizational agility: Risk, uncertainty, and strategy in the innovation economy. *California Management Review, 58*(4), 13–35.

Teece, D., Pisano, G., & Shuen, A. (1997). Dynamic capabilities and strategic management. *Strategic Management Journal, 18*(7), 509–533.

The HP Business Intelligence Maturity Model: Describing the BI Journey. HP. (2007). Retrieved October 30, 2019, from http://download.101com.com/pub/tdwi/Files/BI_Maturity_Model_4AA1_5467ENW.pdf.

Van Beek, D. (2014). *Business intelligence must be more flexible, faster and cheaper*. Retrieved November 22, 2019, from https://www.passionned.com/business-intelligence-must-be-more-flexible-faster-and-cheaper/.

Van Looy, A., Poels, G., & Snoeck, M. (2016). Evaluating business process maturity models. *Journal of the AIS, 18*(6), 1.

Van Oosterhout, M., Waarts, E., & Van Hillegersberg, J. (2006). Change factors requiring agility and implications for IT. *European Journal of Information Systems, 15*(2), 132–145.

Verheyen, G. (2019). *The illusion of agility (what most Agile transformations end up delivering)*. Retrieved October 30, 2019, from https://guntherverheyen.com/2019/01/07/the-illusion-of-agility-what-most-agile-transformations-end-up-delivering/.

Watson, H. J., & Wixom, B. H. (2007). The current state of business intelligence. *Computer, 40*(9), 96–99.

White, D. (2011). *Agile BI: Complementing traditional BI to address the shrinking decision window. Aberdeen Group*. Retrieved December 5, 2019, from https://www.tableau.com/learn/whitepapers/agile-bi-complementing-traditional-bi.

Williams, N., & Thomann, J. (2003). *BI maturity and ROI: How does your organization measure up?* Retrieved December 19, 2019, from http://www.decisionpath.com/docs_downloads/TDWI%20Flash%20-%20BI%20Maturity%20and%20ROI%20110703.pdf.

Zahra, S. A., Sapienza, H. J., & Davidson, P. (2006). Entrepreneurship and dynamic capabilities: A review, model and research Agenda. *Journal of Management Studies, 43*(4), 917–955.

Zimmer, M., Baars, H., & Kemper, H. (2012). The impact of agility requirements on business intelligence architectures. *Proceedings of the 45th Hawaii International Conference on System Sciences (HICSS 45)*, 2012, Maui, HI, USA, p. 4189-4198.

Chapter 9
Business Intelligence Culture

9.1 Human Factors in Business Intelligence

In BI research and professional sources discussing the reasons for business intelligence problems or failures, technological factors are far outweighed by human factors (e.g., Moss & Atre, 2003; Stangarone, 2014). Marchand, Kettinger, and Rollins (2001) have stressed the importance of human factors like information behaviour and values. Presthus (2014) has defined business intelligence systems as socio-technical systems with equal importance given to technical and human factors, and presented a set of business intelligence adaptation cues, based on human factors. Yoon, Ghosh, and Jeong (2014) have stressed social influence and learning climate as important human factors in adopting business intelligence. Villamarin-Garcia and Dias Pinzon (2017) have extracted a list of business intelligence success criteria from the literature, where there are very few technology-related criteria. Yeoh and Coronios (2010), and later Yeoh and Popovic (2015) have defined critical success factors for business intelligence implementation, where organizational and managerial dimensions prevail. Olszak (2016) has explored factors leading to business benefits from business intelligence; again, mostly managerial factors dominate. A concept of organizational absorptive capacity, proposed by Cohen and Levinthal (1990) and discussed in earlier chapters, rests on abilities to merge managerial and informational factors into a capacity to receive and use external information, and encompasses not only the sum of individual capacities, but also the organized ability to exploit them.

However, the studies addressing the role of human factors in business intelligence, and informing in general, are not numerous, and the area is seemingly under-researched. In 2008, Gartner, a consultancy, had rounded up several major challenges in implementing BI projects that are mostly centered around human factors (Table 9.1). One of the important features of these challenges is that many business organizations having undertaken BI projects did not give the human factors

© Springer Nature Switzerland AG 2021
R. Skyrius, *Business Intelligence*, Progress in IS,
https://doi.org/10.1007/978-3-030-67032-0_9

Table 9.1 Principal challenges in implementing BI projects (based on (Gartner Research, 2008))

Challenge	Relation to human factors
The value of BI project is not evident to the business. The project is initiated by the top management, while the other managers and employees, meant to be the core users of BI system, do not fully realize the value that BI might contribute for management and their activities	Users do not relate the BI project to advanced informing and to their actual needs. Projects are isolated; there's no horizontal community
The MS Excel-dominated intelligence still prevails in most companies, where the employees extract data from various operational systems, manipulate it in their personal intelligence space, and produce results for their personal use. Such results are outside reach for the rest of the organization, efforts are often duplicated, different users might produce contradicting results that are hard to justify	Fragmented and uncooperative efforts, and no insight sharing in place
The attention given to data quality is still inadequate. Systems that are based on incomplete, incorrect or doubtful data cannot be used for real management tasks. Such systems and their products experience reduced trust from the users.	Data quality might partly be related to its narrow use in fragmented BI environments, especially external data, and limited data check possibilities. Local sharing-based "crowdsourcing" for data checking might discover possible inconsistencies by including more people and their possibly related data or information. Data governance and stewardship policy is required
A BI system is not a static reporting tool for lifetime use. BI system has to evolve according to changing business needs	The agility of BI system and its users as preparedness for change
To reduce costs and time to operation, some businesses outsource the implementation of BI to an external entity. Often the result of such decision is a BI system that is inflexible and of inferior quality	Outsourced BI is a solution that goes against the very nature of BI as a sensitive strategic function
In many organizations BI implementation faces a lack of common understanding and shared meanings; e.g., even a simple term like "income" might be treated differently by employees of different departments	Common understanding and shared meanings very directly relate to BI culture

required attention, and had ended up with a fragmented implementation of BI or not using the technology at all.

The above set of challenges indicates that, obviously, human-centric problems dominate over the technology-centric ones—BI value to business, fragmented and disjointed BI practices, rigid systems that quickly fall behind the dynamic business context, lack of common understanding and shared meanings are indicated in numerous research in the years following after Gartner's study.

An interesting approach supporting the importance of human factors in BI adoption is presented by Tyson (2006), defining the order of procedures when implementing competitive intelligence. The implementation of an IT-based system should be the last stage of a buildup of competitive intelligence process; however, many companies start exactly from this implementation. While proper intelligence procedures and practices are not implemented in the organization, it is suggested to hold back the wide-range computerization of intelligence processes. To be precise, Tyson has addressed only the issues of competitive intelligence; however, competitive intelligence is the part of BI activities that is the least structured, most turbulent and uncertain, as compared to internal intelligence, and cultural issues are of prime importance in dealing with this uncertainty.

The wide and varied set of human factors, although important, is rather fragmented and lacks a unifying concept. An emerging trend is the development of specific business intelligence culture (BIC) as a collection of attitudes, norms and values, which joins together the human traits of business intelligence and serves as a unifying environment for human drivers and a key prerequisite for successful adoption. The issues of business intelligence culture are still lacking structure and clarity; however, several groups of culture features are starting to emerge. This has led to the attempts to gain insights into key issues related to business intelligence culture by studying existing research and examining the information activities in organizations that already have adopted business intelligence and have considerable experience in its use.

9.2 Relation of Organizational, Information and Business Intelligence Cultures

The set of human factors and related cultural issues, explored in published research, has hinted an existence of cultural environment to join human factors of business intelligence. The terms like information culture, business intelligence culture, intelligence culture, analytic culture, decision-making culture, data-driven culture are already in use by researchers (Kiron & Shockley, 2011; Marchand et al., 2001; Marchand & Peppard, 2013; Popovič, Hackney, Coelho, & Jaklic, 2012; Viviers, Saayman, & Muller, 2005; Wells, 2008). As Höglund (1998) properly noted, culture, like information, is a broad and vague concept that is difficult to define. To avoid confusion between various terms aimed at cultural issues in an organization, the author suggests to clarify the relation of organizational culture to informing-related culture areas along the same three-level structure that has been used discussing BI maturity and agility:

- **Organizational culture (OC)** covers the following dimensions: values, norms, behaviour, habits; information and informing processes are an important part of organizational culture, as stated by numerous sources (Herschel & Jones, 2005; Hough, 2011; Lee & Widener, 2013);
- **Information culture (IC)**, according to Höglund (1998), and Choo (2013), constitutes part of the organizational culture and has its own set of dimensions. It covers all informing processes in an organization from the most detailed data units and actions performed on them to the complex information needs and complicated procedures to meet them. Or, according to Choo (2013), information culture is a part of organizational culture that is concerned with the assumptions, values, and norms that people have about creating, sharing, using information;
- **Business intelligence culture (BIC)** may sound somewhat controversial in a sense that it is a relatively new term, and the understanding of it varies from source to source. Although vaguely defined, BIC may be considered part of information culture, and its positioning is clearly on the complex part of information needs scale, intended to satisfy the needs for insights and decision support that arise from important and possibly costly issues.

The research sources, discussing assorted features of OC, point out the shared value and norms of the collective organization (Lee & Widener, 2013); relationships, effectiveness, alignment, accountability, responsibility, commitment, change, values (Wells, 2008); inclusive systems of participation that promote commitment, and consistent information from recognized leaders signalling what is important (O'Reilly & Chatman, 1996); determination to collect, process, analyze and share information, based on facts and knowledge, trust, HR management with emphasis on analytical and creative people (Olszak, 2016); ability to create a common culture and shared understanding of the environment in which the organization operates (Meredith, Remington, O'Donnell, & Sharma, 2012). Summing up, OC can be seen as a set of norms and beliefs that are valid within the bounds of an organization, significantly related to informing in terms of information access, distribution and decision making. The important role of information creates an overlap between OC and IC; similarly, BIC may be considered a part of IC, although carrying its own specific features. The specific nature of cultural issues arising in business intelligence may be supported by the fact that neither OC nor IC deal with issues like intelligence community, or integration and sharing of intelligence information as their key features. Also, these issues are hardly mentioned when discussing implementation of regular systems of the enterprise resource planning type.

9.3   The Importance of Informing Activities for Organizational Culture

The wide context of BI acceptance and use has attracted research attention to nature of relations between OC and informing activities. Arguing about business culture, Wells (2008) relates its key areas to terms that are encountered in papers dealing with BIC, e.g., analytic culture, culture of discovery, culture of measurement, learning culture, decision-making culture etc. Watkins (2013) provides definitions of OC that expose several traits featured in discussions on BIC: incentives, shared understandings and sense-making, values and rituals, norms and definitions of right and wrong, defensive system that senses risks and should be able to change with the environment. Discussing maturity of business analytics, Cosic, Shanks, and Maynard (2012) have pointed out that culture influences key informing functions: the way decisions are made (gut feeling or fact-based); the use of key performance indicators and quality measurement; the degree to which business analytics merge into daily business activities; the level of management support for business analytics; and receptivity to change. Arguing about BI acceptance, Grublješič and Jaklič (Grublješič & Jaklič, 2015) name OC and IC as important determinants, where IC includes information transparency, openness in reporting and presentation of information on errors and failures. An interesting point in the research of Grublješič and Jaklič is that when asked about IC, all interviewees in presented research have unanimously pointed this determinant as very important to BI acceptance; however, a related question on OC has fared much more modestly. This has raised an interesting side question on how much OC and IC relate, at least in the perception of the interviewees. Grublješič and Jaklič also point out to the importance of social influence if the demonstrable results of using BI positively influence professional image of BI users. Mulani (Mulani, 2013) defines issues-driven culture as an approach making data an asset to the business—right data at the right time and place, displayed in the right visual form, ensuring decision makers get access to intelligence, rather than just more information. Olszak (2016) used a term of "analytical erudition", encompassing flexibility, agility and creative interpersonal communication. In researching the issues of BI acceptance, Popovič et al. (2012) have related BI acceptance to what they have called an analytical decision-making culture (ADMC) that can help with overcoming the well-known tradeoff between reach and richness; a larger number of knowledge workers will use more complex BI systems and more comprehensive information.

Many sources start the discussion on cultural issues not talking about BIC or analytical culture, but instead look at the types of organizational or business culture and relate it to the BI strategy. E.g., Boyer et al. (2010) accentuate the need to understand the business culture of an organization and relate it to the goals of the BI program. By doing this, one can better understand the challenges of BI acceptance both from management and user community, and define the appropriate drivers providing momentum to the program. Evidently, as it has been already mentioned, some types of OC are more accepting for BI, but even in complicated cases where

Table 9.2 Features of information culture and their relation to business intelligence functions (based on (Choo, 2013))

Type of information culture	Features	Dominating business intelligence functions
Risk-taking	Sharing and proactive use of information are emphasized, promoting innovation and insights; focus on external information	Market intelligence and competitive intelligence, often ad-hoc
Relationship-based	Focus on internal information, promoting collaboration and cooperation; again, sharing and proactive use of information are emphasized	Communication, information integration, positioning and strengthening of organization-wide business intelligence function
Rule-following	Efficiency, control and integrity are the key features; focus on internal information	Monitoring of internal status on permanent basis
Result-oriented	Emphasizes evaluation of performance of organization as a unit, focusing mostly on external information and using benchmarking, surveys, market research	Market intelligence and competitive intelligence, largely by defined procedures

the existing OC does not look rather catalytic, the proper understanding of culture factors should reduce the risk of project failure.

Choo (2013) relates OC to IC, and presents two dimensions of the latter: information values and norms, and information behaviour. Alongside these dimensions, four types of IC are specified: risk-taking, relationship-based, rule-following, and result-oriented; and roughly conforming to four OC types proposed by Cameron and Quinn (2011): adhocracy, clan, hierarchy, and market. As business intelligence strives to provide complete view of the environment, where all information sources and processing functions have their important role, the presence of intelligence activities exists in all types of IC proposed by Choo (Table 9.2):

Not all types of OC are equally supportive of business intelligence. There is a need to be aware of certain cultural norms to be avoided as those limit the value of BI. An example would be the culture of caution that might block the sharing of BI insights. In order to gain most value of BI, the culture of transparency and openness is required and in the organization where those values are discouraged, BI won't be working properly since BI implementation might discover difficult truths about the organization. Herschel and Jones (2005) have concentrated on the factors limiting information and insight sharing, and state that insight sharing is not supported in cultures where people are rewarded for what they know and others do not. Brijs (2013) has stated that the machine bureaucracy is incompatible with sophisticated business intelligence and knowledge exchange for strategy development. According to Parr Rud (2009), "... Sadly, the culture in many organizations dictates that mistakes are bad and should be avoided at all costs. Employees are criticized or even ridiculed for mistakes. This stems from early learning in institutions where mistakes meant a lower grade and even possible consequences at home." Discovering insights and responding needs a "degree of nimbleness that can't be achieved by hierarchical bureaucracies" (Appelbaum, Calla, Desautels, & Hasan, 2017). To work around the rigidity of hierarchical structures and centralized decision making, alternative organizational structures like horizontal networks of subject matter experts are required—flat and fluid structures that allow horizontal flow of information and ideas, and are much more catalyzing for richer meanings, faster and better decisions.

It has been stated in the previous chapter that the level of organizational, information and BI agility is largely dependent on cultural factors. The misfit between technology and culture may initiate employee resistance and impair BI agility. Boyer et al. (2010) point out to several sources of employee resistance due to cultural factors:

- They find that changing their habits—those age-old, familiar ways of doing things—is challenging, and the value may not be understood
- Sometimes, they worry about accountability and believe that a change in the familiar processes of their jobs will uncover something they would rather not reveal.
- They resist because they feel they were not consulted in the process—perhaps because they had needs that were different from those of the rest of the organization.
- They don't understand how the new processes that will be implemented can benefit the overall organization, or they fear that the change may dilute the value that they are already providing to the organization.

The above sources of employee resistance are based on natural fears; however, they all have a cultural background—attitudes to change, mistakes, revealed weaknesses, neglect of individual needs, failure to communicate the big picture and related benefits.

The term "intelligence culture", although not exactly obscure, is still somewhat vague and has been assigned multiple meanings. On the other hand, many sources agree that BIC is a part of the wider concept of organization culture. In discussing human factors of BI adoption, many sources refer to culture issues, assigning different contexts for culture. The different aspects of BI culture issues and the suggested features of their context are presented in the Table 9.3 below.

Summing up the presented culture issues, we may note that there are common features across all presented definitions of BI culture or related concepts. Most of them point to an environment that is inclusive and democratic; important information processes are understood and supported; the principal drive for leveraging BI potential comes from the users and management, not the availability of technology. This allows the more exact definition of BI culture: BI culture is part of organizational culture, and represents a set of norms, rules, attitudes, and values that acts as a catalyst in creating value by providing actionable insights to decision makers.

Table 9.3 Features of various types of culture issues in BI

Source	Aspect of culture	Features
Grublješič and Jaklič (2015)	BI culture	Committed management support Adequate organizational culture Adequate information culture Appropriate change management practices
Kiron and Shockley (2011)	Data-oriented culture	Analytics used as a strategic asset Managerial support for analytics throughout the organization Insights are widely available to those who need them
McAfee and Brynjolfsson (2012)	Decision making culture	Leadership Talent management Technology Decision making Company culture
Popovič et al. (2012)	Analytical decision-making culture	A decision making process exists and is understood Available information is considered regardless of decision type Such information is used for each decision process Shared use of BI functions and information
Mulani (2013)	Issues-driven approach	Questions address the business issue, not the data Known needs of users New tools and techniques for information integration and visualization
Harris (2012)	Analytical culture	Employees are good at creating questions and working collaboratively with the data scientists
Presthus (2014)	Information infrastructure	Concentration on human factors of technology adoption Simple and useful tools whose benefits (faster data access, easier analytical functions) have been communicated clearly from the beginning The IT platform stimulates use by being simple, shared and open Self-reinforcing installed base—when users contribute, the user base and the value of system increases Improvements and additions are easy and performed in-house Expandable with universal standards, preventing eventual lock-in

Source: author

From the above analysis of forces driving BI and features of BI culture issues, several groups of factors that are most important in influencing BI culture and contribute most to the value created by BI investments are presented in the next paragraph. These groups are: sharing and cross-functional BI activity; record of lessons and experience; creation and presence of intelligence community; supportive technology management.

9.4 The Most Common Features of Business Intelligence Culture

Although the published research sources on business intelligence culture are not very numerous, their analysis has produced a set of the most common features of business intelligence culture, named below.

9.4.1 The Importance of Information Sharing

While for a lone analyst the importance of the skills to cope with intelligence tasks is obvious, many sources have stressed that communication between participants in the business intelligence process should create value by motivating tool adoption and use or insight development. This involves hierarchically horizontal information transfer (peer-to-peer), as opposed to the involuntary information reporting structures based on vertical information transfer (Davenport, 1997). It has to be noted that in literature terms "information sharing" and "insight sharing" are used interchangeably. To clarify this issue, we will consider insights as a specific type of information, loaded with aggregated and valuable sense, and use a single term "information sharing" to denote both sharing of information and insights. Carlo, Lyytinen, and Boland Jr (2012) defined the existence of *collective mindfulness* as a set of combined activities that create awareness and facilitate discovery in high risk environments. Davies (2004), arguing about intelligence culture, has stressed the importance of communication and sharing of information:

> The need for collegiality ... is driven by specific conditions. Those conditions include technically and technologically diverse intelligence sources, the interdependence of those sources as pieces of a larger picture, and the need to meld those pieces together to get a coherent picture on the basis of which decision-makers can make their decisions

As stated in Chap. 2, BI has many relating points to intelligence activities from other areas, and cultural issues have common traits as well, including information sharing. Regarding the events of September 11, 2001, as one of the worst cases of inadequate intelligence information sharing, we can quote Hamrah (2013):

> Overall, the culture of the Intelligence Community does not foster collaboration and intelligence sharing. This was evident when analysis was not pooled and effective operations were not launched to prevent the 9/11 attacks. ... Each intelligence agency has its own core competencies, systems and data warehouses, which makes it difficult to piece together factors that together could help provide warning of future threats and give us grater insight regarding our adversaries. However, turf wars, stove piping or the simple inability to share information continues to hinder effective information sharing and the Community's ability to put pieces of the puzzle together as effectively and efficiently as possible.

The opening of information silos inside the organization and horizontal cross-function exchange of information is one of the most prominent features of business intelligence culture found in published research. The horizontal nature of BI function and the capability of BI to cross-functional borders and join data and information buried in functional silos has been stressed by almost every source on BI adoption. Presthus, using the title "information infrastructure", which is more likely to represent BI architecture, has stated that it unites technology, people, processes, communities, and tends to be self-organizing (Presthus, 2014). Kiron and Shockley (2011), describing elements of data-oriented culture, have stressed the sharing of information as one of the key characteristics of this culture. They also stated that BI insights should be widely available to those who need them, and presented an example of BT (formerly British Telecom) that had linked its data silos to dramatically improve its customer service. Marchand and Peppard (2013) suggest to assemble cross-functional analytical teams, open data silos, and facilitate a culture of information sharing. Matzler, Strobl, and Bailom (2016) use the term collective intelligence and point to the importance of creating cognitive diversity with more contacts outside the usual circles, and the ability to access decentralized knowledge. Hackatorn and Margolis (2016), discussing immersive analytics, noted that persons within a collaborative environment generalize beyond known data and identify situations where human judgment is required. Choo (2015: 149) noted that in organizational information seeking, other people are most important sources of information, and are preferred when context and hidden aspects are no less necessary than facts. Organizations with strongly compartmentalized work, and attention as a scarce resource, offer little incentive for additional information seeking across compartment borders. Presthus (2014) describes actual experience from implementation of a BI system, and has pointed out that the information technology platform will stimulate use if being simple, shared and open. A good example of relation between OC and information sharing, already mentioned earlier in this book, is Xerox Eureka, technically a collection of tips for copier maintenance that grew into expertise-sharing platform (Cronin, 2004):

> . . . Eureka is a database of tips on photocopy repair, created and used by repair engineers. In the sense, it represents and expertise database for front-line service engineers. Although the virtual community that fromed around the development o Eureka is not, strictly speaking, a community of practice, Eureka is a forum for the sharing of expertise similar to a community of practice. . . . Before Eureka there existed a more localized information-sharing culture. Eureka's success is due, in large part, to to the perceived quality of the system and to the pre-existing organizational culture and behavior of the engineers. Awareness of the organizational culture and the information behavior of the engineers is key to understanding the success of Eureka in knowledge management rather than in IT terms. For collaboration to take place, an appropriate organizational culture must exist or be fostered. This may be described as an information-sharing culture, and the importance of this type of culture to knowledge management is reflected in the interest the knowledge management culture takes in organizational culture. Creating an information-sharing culture within an organization can be difficult.

Although the above example is presented in the context of knowledge management, it rather centers on information sharing and its relation to organizational culture. Mettler and Winter (2016) have researched determinants of information sharing in organizations by using social network platforms to exchange any information. They tested assumptions about relationships between information sharing in enterprise

Table 9.4 Most often quoted BI functions (IT Summit, 2015)

Functions provided by a BI system	Share (%)
Data security and access management	93
Real-time analysis	86
Data collection from many sources and systems	84
Possibility to share created reports and insights	84
Ad-hoc analytics for self-serving users	77
Data drill-down	71
Predictive analytics	68
Intuitive and clear user interface, requiring no training	58
Use of mobile devices	55
Data mining options	46

social systems and such determinants as organizational information ownership norms (not significant), reciprocity and social cohesion (significant), quality of shared information (significant), and privacy concerns (significant).

For data-driven approaches, BI participants share resources, while for the needs-driven approaches they share the results. The experience of intelligence activities in areas other than business (Johnston, 2005: 67–68) also supports creation of additional value by using additional expertise of other participators. The importance of sharing is reflected in the empirical data from a survey of 69 business managers performed by IT Summit, a consultancy (2015) in Table 9.4, showing the most often quoted functions of a BI system in use.

We may note that functions supporting sharing of created reports and insights have been indicated by 84 percent of the surveyed business managers. The same research has shown that BI users who have practical experience in using BI systems rate sharing functions higher than non-users: 52% of users rated sharing options between 9 and 10 points (out of 10), as compared to 40% of non-users for the same range of ratings.

The issues of motivation of information sharing have been researched by March-and et al. (2001), who listed behaviors and values that enhance integrated information use, including transparency as an approach to deal with mistakes, errors and failures positively, motivating organizational members to share important information. They also stated that information may be more freely shared among individuals or small teams than between departments in a company, thus supporting the point that these individuals or small teams actually constitute the intelligence community inside an organization. Uploading of information to be shared by a member of intelligence community can be motivated by former reception of useful information and mutuality. A cooperative and catalytic business intelligence environment might utilize the filtering potential that information sharing is expected to have on information integration, and have a positive effect on objective reflection of reality that business intelligence provides. The validation aspect has been supported by Elliott (2014), who noted that local analytics used for decision making can be validated through online sharing and review, or ‚community policing'.

Sharing of information is not without limits and invokes certain risks: as importance of shared information grows, so might its sensitivity together with protection issues. Cases with confidentiality or risk of abuse require additional access control that is best controlled by information owners (Denning, 2006). The sharing of information may be blocked by the culture of caution, as integrated meanings might discover inconvenient issues about the organization. There also is a possibility that influenctial alternative groups might form and manipulate the shared information. The risk aspects of information sharing are not discussed in this book, and, to author's opinion, might deserve a separate research.

9.4.2   *Record of Lessons and Experience*

The role of lessons learned and experience is discussed in detail in Chap. 6 of this book. As suggested by the importance of sharing of information in the previous paragraph, the horizontal nature of the business intelligence function is also supported by collections of accumulated and shared previous experience and lessons. These collections include not only success stories, but errors, failures, mistakes and surprises, thus reinforcing trust among the members of business intelligence community (Marchand et al., 2001). Grublješič and Jaklič (2015) indicate availability of prior experience as one of the business intelligence acceptance factors. Discussing expertise management in an organization, Shah (2011) stresses the need to use social systems that aggregate available experience and develop competitive advantage through driving collective talent. Hough (2011), discussing ways for business intelligence function to support organizational strategy, suggests identifying, sharing and evangelizing best practice; and sharing and discussing assumptions and strategies with everyone involved. Kiron and Shockley (2011) point out that the build-up of analytics expertise and lessons learned is an important link in the information integration chain.

Whatever system is proposed for LL environment, is has to be user-driven and promote voluntary collaboration as vital part of the analytical culture; many attempts to implement a centralized scheme have failed. It also relates to benefits and risks—analytical culture should provide benefits outweighing the risks. The benefits of using this system/environment should be simple and clear; the future benefits may show up in unexpected places as derivatives of the value potential. The risks of participating and sharing like job security risks might prevent contributors from contributing valuable professional information to LL; breaches of confidentiality and risk of abuse have to be accounted for as well.

9.4.3 Creation of an Intelligence Community

Many researchers agree that people actively engaged in business intelligence activities tend to gravitate towards forming a business intelligence community. Hallikainen, Merisalo-Rantanen, Syvaniemi, and Marjanovic (2012) indicate that to propagate business intelligence thinking, a shadow community, where mental models and outcomes are shared, is of paramount importance. According to Mettler and Winter (2016), regular interaction on a sharing platform strengthens social ties and may lead to social cohesion, while peer pressure and increased visibility motivates individuals to maintain high quality of shared information. Alavi, Kayworth, and Leidner (2006) point out that the bonding glue for communities is shared professional interest and commitment, and informal communities are more agile and resistant to change than formal ones. Presthus (2014) has stressed the importance of self-reinforcing installed base: when users contribute, the user base and the value of system increases. An interesting point raised by Yoon et al. (2014) suggests that perceived social influence from referent others like co-workers or supervisors, or a learning climate, have a significant positive influence on individual intent to adopt business intelligence applications. Marchand and Peppard (2013) recommend finding sources of positive cycle to help reinforce the business intelligence community, and provide a case study of business intelligence community growth with positive feedback. Marchand et al. (2001) present a case where people have competed with each other in skills of using a business intelligence system and speed to locate the necessary information.

The intelligence community—a set of people in an organization that execute intelligence should be sustainable and self-reinforcing. It should comprise analysts and insight builders from key functional areas, and with different analytical background; according to Harris (2012), data scientists should work collaboratively with business users that are good at asking business questions. To avoid becoming a counter-culture, this community should possess leadership and serve as change agents having the mandate of management to drive BI culture in an organization. This leadership, according to (McAfee & Brynjolfsson, 2012), should set goals, define success directions and ask the right questions. Gradually the former shadow community should move to the key influencers, or middle managers (Hallikainen et al., 2012).

An important feature for business intelligence community, mentioned in published research, is tolerance for experimentation and mistakes. Marchand and Peppard (2013) stress that building a collaborative culture requires being tolerant for experimenting, learning and errors, and intolerance for failures and errors work against this culture.

Imhoff and Pettit (2004) have proposed a classification of user types that exist in intra-organizational communities:

1. Farmers—customers of reporting, e.g., sales and product analysts; well-defined and consistent requirements; repeating informing procedures.

2. Tourists—broad business perspective to assess the general health of the company; impatient and hard to please; need dashboards and quick queries.
3. Operators—clerical staff with tactical focus on today's problems; candidates for procedure automation.
4. Explorers—insight builders and "out-of-the-box" thinkers; operate on intuition and observation and in heuristic manner; create assumptions and look for patterns or ask miners to test the assumptions—actually generate deep business questions; misses are common but hits may have tremendous rewards; the predecessors of "data scientists" with emphasis on insight building.
5. Miners—equivalent of Big Data analysts; "often have a very good idea of what they expect"—are good at formulation business questions.

Boyer et al. (2010) define communities as structures that brings together the BI team members, IT professionals, business analysts, and the larger community of stakeholders who will work closely to achieve success. This technique accomplishes a number of things:

- **Education**—creating a team can more fully educate the participants in the overall BI initiative, helping to identify how individual job functions contribute as elements within the structure of the BI process.
- **Structure Support**—the creation of operational support for the decision making and execution.
- **Obstacle Detection**—a team, working to define and describe individual business problems, can operate as a mechanism to reveal real-world operational obstacles early.
- **Collaboration**—various business teams working together have permitted valuable cross-pollination of new ideas.
- **Buy-in**—teaming business unit stakeholders with BI professionals creates the opportunity for buy-in by the team members who were initially reluctant; communicates success within the organization; and creates the potential to become a self-reinforcing mechanism for the entire project. That success is virally broadcast to other areas in the corporation.

Franklin, Halevy, & Maier (2005), discussing communities' role in culture, noted that social structures enable groups of people to share knowledge and resources in support of collaborative activities. Such structures should bring together people who each know something but do not know other things—a "symmetry of ignorance", resulting in the exploitation of synergy and "wisdom of crowds". Several hypotheses raised by Yoon et al. (2014) suggest that perceived social influence from referent others like coworkers or supervisors, or a learning climate, have a significant positive influence on individual intent to adopt BI application. This statement suggests the existence of self-supportive BI culture, of which supervisors, peers and learning climate clearly are part of. Self-reinforcement mechanism is discussed in (Presthus, 2014): the key concepts are the installed base (number of participants) and a cycle driven by positive feedback. Once the initial group of participants delivers positive results, it gains credibility and value to organization, causing additional people to join the group.

9.4.4   Technology Management

The technology advances are probably the best-seen signs of BI progress, yet their combination with under-developed human issues has led to disappointments and low BI acceptance. The role of technology, as seen from the discussions on BI agility and culture, is to remove obstacles in utilizing the powers of IT and support the possible reengineering of information relations. Perhaps the dominating factor regarding the role of IT in developing intelligence culture is to provide simple-to-use tools and techniques, where the users may spend more time concentrating on gaining insights and solving problems than on mastering the technology. This point is supported in (Presthus, 2014), where a case of developing and reinforcing the BI community ("installed base") is presented; both attraction of initial base of users and its expansion are boosted by simple and easy-to-use IT capabilities. Barton and Court (2012) have presented cases where the use of a simple tool to deliver complex analytics substantially improved business processes. Yoon et al. (2014) have shown that complexity of BI applications negatively affects BI adoption. Similar results have been shown in (Grublješič & Jaklič, 2015).

An important technology issue is the placement of BI function in an organization, deciding between centralized and decentralized approaches. Cohen and Levinthal (Cohen & Levinthal, 1990), discussing the organization's absorptive capacity, had pointed out that an important feature of this absorptive capacity is information transfer across and within subunits, and difficulty may emerge under conditions of rapid and uncertain technical change if an environment monitoring function is centralized. When external information of random nature is received, and it is not clear where in the firm or subunit a piece of outside knowledge is best applied, a centralized gatekeeper may not provide an effective link to the environment. Following these arguments, centralized approaches for BI are not agile enough and may effectively limit the potential of intelligence function; this point indirectly confirms the potential agility of **decentralized** and horizontal approaches to the distribution and flows of intelligence information as an important feature of intelligence culture. This point is also, if indirectly, supported by research performed by IT Summit, a consultancy, that surveyed 69 business managers—actual BI users on the issues of BI adoption (IT Summit, 2015):

- When asked about dominating sources of decision support information, the largest share, or 68% of business managers pointed to analytical information produced by users themselves from BI systems;
- When asked about BI service latency, the largest share, or 72% of business managers, indicated that they may produce intelligence information (consolidated reports) by self-service anytime.

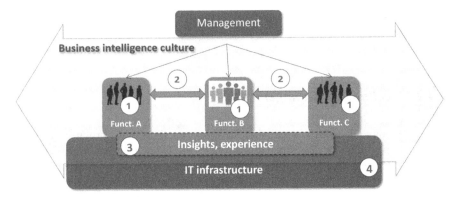

Fig. 9.1 The positioning of business intelligence culture in an organization

9.4.5   The Environment of Business Intelligence Culture

Summarizing the results of the literature analysis by the most prominent features of business intelligence culture., the environment of business intelligence culture might include the participants and processes shown in Fig. 9.1.

In Fig. 9.1, the numbers refer to the corresponding business intelligence culture components listed below:

1. The horizontal intelligence community, comprising analysts and insight builders from key functional areas.
2. Sharing of intelligence information; synergy in a community leading to integration of sense-making and skills.
3. Structured collections of insights, experience and lessons learned; with growing contribution, community size and the value of the system increases.
4. Balanced information technology management, blending centralized elements implying few essential common standards with distributed local and self-service environments that carry the required agility and flexibility.
5. The role of management, although not explicitly pointed out, is to create and maintain a blend of the above components to ensure development of business information culture and business intelligence culture as its subset.

9.5   Research on Business Intelligence Adoption In Lithuanian Companies

This part of the chapter reflects a part of ongoing research into dimensions of human factors of business intelligence implementation, mostly concentrating on user ratings of business intelligence culture in their organizations (Skyrius et al., 2018). The important part of research is the choice of perceived level of business intelligence culture, as seen by the respondents, for the role of a reference point in an attempt to estimate important factors relating to this perceived level. The analysis of published sources on cultural issues around business intelligence has driven a need to perform empirical research in Lithuanian businesses—active users of business intelligence systems to estimate the role of human factors and business intelligence culture in the findings. Business intelligence culture is a relatively new research area, hence to discover important emerging issues it has been decided to employ elements of both quantitative and qualitative methods. The use of qualitative methods containing inductive approach and discovery techniques by eliciting free-text responses are deemed suitable for discovery of emerging important issues. Although not used extensively in this research, the qualitative elements are expected to facilitate the understanding of individual perceptions. The structured part of the survey uses quantitative methods for evaluation of awareness of business intelligence cultural issues among respondents. Because of the unfamiliarity of the researched area, the quantitative analysis has mostly used descriptive methods, together with initial statistical testing of factor significance.

The used questionnaire has touched many aspects of business intelligence implementation and use, including issues related to information sharing, and for this part of the questionnaire the **research question** has been formulated as follows: How do the respondents relate their perceived levels of business intelligence culture in their organizations to information sharing?

The research of business intelligence adoption issues in Lithuania was conducted in September–October of 2016. The research target group may be considered a convenience sample as it included Lithuanian companies and organizations that had acknowledged successful recent implementation of business intelligence, and have revenue of more than five million Euro a year. The survey has been conducted online and questioned 207 respondents composed of first and second-level managers responsible for different business areas. 43% of respondents belong to large (in terms of business scale in Lithuania) organizations with more than 20 million Euro annual revenue; 48% of respondents represented companies with more than 250 employees, and in several cases responses were provided by more than one manager from a company. For the purposes of this paper, the actual use of business intelligence, business analytics tools and techniques comes under a single business intelligence term, unifying the monitoring and analytic parts of related activities.

The questionnaire included several groups of questions:

- Respondent demographics: profiles of companies, respondents and business intelligence technologies in use;
- Expectations from business intelligence use against actual results: value created by use, solved and unsolved problems, factors influencing implementation and use;
- Actual experiences in using business intelligence: actual latency of information presentation and its movement in a process chain; access to and sharing of assorted information for decision support; perception and evaluation of business intelligence culture.

The size distribution of the surveyed companies in terms of number of employees is provided below (Table 9.5):

The distribution of the surveyed companies by industry is (Table 9.6):

Regarding the use of business intelligence tools, all the respondents have already implemented and are actively using different business intelligence software platforms and tools, counting one installation per one organization (Table 9.7):

The results of the research confirmed once again that Microsoft Excel is the dominating business intelligence tool, with 51% of respondents (lines 1, 2 and 4) using Excel or proprietary software developed for a specific company, usually based

Table 9.5 Size of the surveyed companies in number of employees (Skyrius et al., 2018)

Up to 50 employees	32
Between 50 and 250 employees	76
More than 250 employees	99

Table 9.6 Surveyed companies by industry (Skyrius et al., 2018)

Manufacturing	59
Commerce	57
Services	30
Public sector	21
Telecommunications/IT	16
Finance/insurance	10
Other	14

Table 9.7 Use of business intelligence tools (Skyrius et al., 2018)

No	Software platform	Units	Share (%)
1	Unique software developed for a specific company	49	24
2	Microsoft Excel	38	18
3	Two or more different business intelligence systems (Business Objects (www.businessobjects.com), IBM Cognos (www.ibm.com), Oracle (www.oracle.com), Qlik (www.qlik.com) etc.)	34	16
4	MS Excel and proprietary software developed for a specific company	19	9
5	Qlik	17	8
6	Business Objects	15	7
7	Excel and business intelligence system (Business Objects, Cognos, Oracle, Qlik etc.)	12	6
8	Oracle	9	4
9	Power BI	4	2
10	Cognos	3	1
11	Web Focus)	3	1
12	SAS	2	1
13	IBM SPSS	2	1
	Total:	**207**	**100**

on Excel as the only business intelligence tool companywide. On one hand, easily available business intelligence tools, e.g., systems based on Excel, have a significant presence. On the other hand, there is a significant level of adoption of advanced business intelligence technologies in the surveyed sample—49% of respondents (all other cases) are using specialised systems.

Factors Hindering Business Intelligence Implementation
Articulating the factors that hinder business intelligence implementation, the respondents mostly pointed out:

- Lack of BI strategy (29%);
- Lack of BI ambassador—not a project manager, but an internal person who has authority and power, believes in business intelligence and spreads the mission to the whole company (28%);
- Internal disagreements, dilemma of the owner (27%);

Sharing of Information
The research has shown that 60% of surveyed organizations using business intelligence successfully solved the issue of the lack of up-to-date information, and the use of business intelligence allowed 52% of surveyed companies to create an integral system ensuring a "single truth" version of analytical information. A rather high figure, 93% of surveyed organizations indicated they are sharing the information. Depending on the needs, the employees share their information communicating with each other permanently and initiating this process (51%) by themselves or sharing the information during joint meetings (42%).

The use of a business intelligence platform's functionality, for example, to collect data from a large number of different internal data sources (*the zoo of information systems*) to one integral system created significant value in 61% of companies. Besides technical features, it is important to note that in many cases business intelligence has become the main axis of all the organizational processes. It naturally forces the organization to combine all of its activities, monitor and integrate the data of different business areas in a wide overall context. Business intelligence culture actually can be the starting point for a companies' evolution from functional to process-based, supporting a more effective model of doing business in highly competitive environment.

58% of respondents stated that high business value was created by business intelligence systems enabling the sharing of created reports as well as the opportunity to prepare the insights centrally and to present them in advance. At the same time, the developers of business intelligence state that the collaboration functionality of specialized BI systems, providing automatic or automated sharing of prepared reports, is still not widely used. When these responses were processed in the context of the answers to other questions, it became clear that this is more about sharing information by all possible channels, not exclusively using collaboration features of business intelligence system.

Another issue related to potential value of sharing information has been shown by 48% of managers stating that they have to see the context in order to perform own analysis, requiring the information from other departments or processes. However, one-third of the managers use the context only when something is going wrong in their department, and thus the problems are solved post-factum, or just recognized and considered being *lessons for the future*. Only 35% of the respondents were able to access the needed information by themselves; the other 65% had to ask for it from other sources—another department or analysts.

The respondents have been asked to indicate their perceived level of business intelligence culture in their organizations on a five-point Likert scale, value 1 being the lowest rating, and 5 the highest. The most important issue for data analysis below is the features of business intelligence system for the levels of perceived culture equal 3 and more, i.e. average or above. In the following analysis, the indicated level of culture has been related to several questions related to information sharing, and the empiric data have been processed using IBM SPSS Modeler® 17.1 (https://www.ibm.com/analytics/spss-statistics-software) data mining software package to indicate the most frequent relations. The relevant questions are coded as follows:

- Q11—How business intelligence reports are prepared?

  - Self-service on one's own;
  - Reports are renewed automatically;
  - By analysts or IT people on request;
  - Other.

- Q12—In what cases analysis requires information from other processes or departments?

  - In all cases;
  - Only when a problem occurs;
  - Does not require;
  - Do not know.

- Q13—How do you get the decision information from other departments?

  - On my own, having access rights;
  - By making a request to other department;
  - By approaching a person responsible for business intelligence (a business intelligence ambassador);
  - This process is unregulated and chaotic;
  - There is no such need.

- Q15—How do people share analytical insights inside organization?

  - Do not share;
  - Only during common meetings;
  - On permanent basis by direct communication and self-initiated discussions;
  - Do not know;
  - Other.

Table 9.8 Distribution of responses about perceived level of business intelligence culture (Skyrius et al., 2018)

Perceived level of business intelligence culture	Number of cases
1	3
2	18
3	86
4	79
5	21

The role of the dependent variable has been assigned to the perceived level of business intelligence culture inside an organization, indicated in responses to question 23 (Q23) and varying from 1 (poor) to 5 (excellent).

The distribution of responses to Q23 is presented in Table 9.8 below. We have to note that the absolute majority of respondents (186 out of 207, or 89.9%) consider their organizations having their level of intelligence culture at 3 (mediocre), 4 (good) or 5 (excellent).

The respondents were asked to provide a free text comment explaining their rating. For the lowest perceived level of business intelligence culture, equal to 1, just three instances are available with no text comments except one: "Information is being hidden". For other levels, the text comments provide some insights on the respondents' rating of their business intelligence culture. For an indicated weak business intelligence culture level (rated 2), the example comments are:

Fragmented and does not encompass all aspects of activities. The system is still in the launch phase, inexperienced users; hope that culture will evolve. No self-service analytics, just prefabricated templates are used. Lack of strategy and professional analysts. The management does not see business intelligence as an important function. Different organization units see things differently.

It can be noted that for the level 2 of business intelligence culture dominating response topics are: business intelligence is in its initial stages of implementation, and the importance of business intelligence and its use are quite low; business intelligence is fragmented and chaotic, there's no unified understanding for different units, and many obstacles and limitations lead to low usage.

For indicated mediocre business intelligence culture level (rated 3), the example comments are:

The required level of implementation, management and use is not achieved yet. Although the organization is a rather advanced user of business intelligence, processes are still upgraded because there's still space to grow. We have all the contemporary tools we need; it's more the lack of culture and methodology. The emergence of business intelligence culture is rather recent, and not everybody in the organization equally understands its importance. Lack of clear intelligence strategy. What are the goals and what are we trying to achieve? We have to master internal communication. There's also lack of competence in some areas.

Dominating response topics point out that business intelligence still inadequate; there's space for growth. On the other hand, respondents are aware of the growth of business intelligence importance and related culture changes. Different people have different competences and attitudes, and there's noticeable lack of competence and support.

For indicated good business intelligence culture level (rated 4), the example comments are:

> There's an attitude that business intelligence is a must, and competence gaps are insignificant. The constantly changing environment forces adapting to changes ... follows one of the quality management principles—admit mistakes. The analysis is directed to the needs of direct users, not top management. Human factors define processes and motivation. Recognition by top management. Data is available, but not convenient—part of information is not integrated, automation is insufficient and requires additional work and time. There's space to grow. We are strong in intelligence, but it can be done better.

At this level of perceived business intelligence culture (4), dominating response topics point out that there is a perception of significant business intelligence role, growing understanding of business intelligence importance, and growing user base. Business intelligence culture seems quite adequate and draws realistic evaluations—nothing is perfect, but the organization steadily moves to commonplace business intelligence. The respondents are also aware of business intelligence value, its strengths and weaknesses, and directions for improvement.

For indicated excellent business intelligence culture level (rated 5), the example comments are:

> Because one can note a serious attitude and processes, understanding, investment into upgrades, inclusion of new related products and additional departments. There are all conditions for working and perfecting, submitting proposals, use required innovations. Business data for analytics are accessible to every specialist inside competence boundaries. Inside the same boundaries, and based on this data, decision rights are granted. Work is hardly imaginable without (business intelligence) tools—decision making would be complicated; employees use them without additional directions. Business intelligence is already in use for some time, there are qualified people and everything works.

At this level of perceived business intelligence culture (5), dominating response topics point out that business intelligence function is clear and well-positioned, and business intelligence activity is naturally woven into work processes. There is a significant body of experience and traditions.

The most mature answers relate to 4 level of perceived business intelligence culture; evidently, the respondents have a realistic estimation of business intelligence culture in their organizations. Meanwhile, the responses with level 5 sound a bit too optimistic. Most of the respondents have an adequate understanding of business intelligence culture and do not hesitate to relate its perceived level to business intelligence implementation success.

To gain more insight into relation between questions in information sharing and perceived level of business intelligence culture, a neural network analysis has been performed on corresponding data using IBM SPSS Modeler® 17.1 (https://www.ibm.com/analytics/spss-statistics-software) data mining software package. The analysis procedure, including questions Q11, Q12, Q13 and Q15 as predictor variables for Q23, had generated a network with single hidden layer containing a single node. The results of analysis are presented in Table 9.9.

Table 9.9 The results of neural network analysis for prediction of Q23—The perceived level of business intelligence culture (Skyrius et al., 2018)

Sample	Observed	Predicted					Percent correct (%)
		1	2	3	4	5	
Training	1	0	0	1	0	0	0.0
	2	0	0	12	3	0	0.0
	3	0	0	38	22	0	63.3
	4	0	0	20	36	0	64.3
	5	0	0	5	8	0	0.0
	Overall (%)	0.0	0.0	52.4	47.6	0.0	51.0
Testing	1	0	0	2	0	0	0.0
	2	0	0	3	0	0	0.0
	3	0	0	15	10	0	60.0
	4	0	0	12	11	0	47.8
	5	0	0	4	4	0	0.0
	Overall (%)	0.0	0.0	59.0	41.0	0.0	42.6

Below we present the more detailed analysis of responses to questions Q12 and Q15

The results indicate that both for training and testing stages of neural network analysis the strongest prediction levels lie in the area of Q23 values of 3 (mediocre) and 4 (good). This can be explained by the domination of these values in survey responses. The overall prediction strength is lower—51.0% for training stage, and 42.6% for testing stage; this setback can be explained by lower frequencies of other instances for Q23.

Q12. Information Needs Outside Own Department or Function

The responses to Q12, asking in what cases analysis requires information from other processes or departments, have been coded as follows: NE—unaware; IN—information from other places is not required; NU—information is required on a permanent basis; IR—information is required only if there is a specific need to determine the cause of a problem, e.g., why some process is stuck.

The most frequent relations in this pair are shown in Table 9.10:

The data show that both for levels of perceived business intelligence culture 3 and 4 in question Q23, the dominating case of answers to question Q12 is "Information is

Table 9.10 Features of information needs outside own department or function (Skyrius et al., 2018)

Q12—in what cases analysis requires information from other processes or departments	Q23	Number of occurrences
NU (information is required on a permanent basis)	4	43
NU (information is required on a permanent basis)	3	38
IR (required only if there is a specific need)	3	33
IR (required only if there is a specific need)	4	26

required on a permanent basis", and points to a perceived need of shared information, reflecting the permanent nature of intelligence and analytical functions.

Q15. Sharing of Analytical Information Inside an Organization

The responses to Q15, asking how analytical information is shared inside organization, have been coded as follows: NK—no exchange of information; KN—permanent exchange by direct communication; KB—information is exchanged only during common meetings; NE—unaware.

The most frequent relations in this pair are in Table 9.11:

The data show that the most common occurrences involve responses "permanent exchange by direct communication" (KN), and "information is exchanged only during common meetings" (KB). In both cases, responses point to active and permanent sharing of information, indicating that the data silos between departments and functions have been opened or did not exist at all.

There is a noticeable relation between perceived level of business intelligence culture and intensity of information sharing between business functions, although its statistical significance is not strong. It can also be noted that, although the need for information sharing has not been very obvious in open text responses to the question Q23 about perceived level of business intelligence culture, the further analysis of data on information sharing for levels 3 (mediocre) and 4 (good) confirms its importance for above-average business intelligence culture.

The set of culture components, drawn from published sources, i.e., information sharing and integration, record of lessons and experience, existence of business intelligence community, has been recognized as important for the formation of business intelligence culture in the responses of business users. The attempt to relate these components to the perceived levels of business intelligence culture may be considered a novel result, unifying together scattered human factors of business intelligence success that were found in literature analysis. An important part of the findings is that the respondents have shown an understanding of the concept of business intelligence culture, of its strength or weakness, and of its role in creating benefits from using business intelligence. These findings justify the distinction of business intelligence culture as a separate concept that unifies the scattered definitions of cultural issues relating to business intelligence in published research. Although this research did not cover relating organizational culture to business intelligence culture, the responses reflect the influence of the overall organizational culture on how the business intelligence potential is utilized. This finding could be a

Table 9.11 Sharing of analytical information (Skyrius et al., 2018)

Q15	Q23	Number of occurrences
KN—permanent exchange by direct communication	4	47
KB—information is exchanged only during common meetings	3	40
KN—permanent exchange by direct communication	3	36
KB—information is exchanged only during common meetings	4	29

direction for future research avenues concentrating on relations between organizational culture, information culture, and business intelligence culture.

The analysis of empirical research data has shown that in the surveyed organizations the dominating share of respondents perceived their business intelligence culture as being mediocre to excellent. As the results deal mostly with intelligence information sharing, it can be noted that the surveyed organizations are well aware of their analytical needs and are quite active in supporting the horizontal communication of shared information. For good and excellent levels of perceived business intelligence culture respondents have indicated permanent needs and use of information from other functional departments, and active exchange of insights by direct communication. The importance of information sharing, as pointed out in many published works dedicated to the human factors of business intelligence implementation, has been only partially confirmed by the results of empirical research. The modest statistical significance of the relationships between the variables introduces reasonable caution in producing conclusions. On the other hand, respondents who are the most satisfied with their business intelligence culture seem to exist in an environment where information sharing is a natural way of executing business intelligence, and data silos between departments and functions have been opened or did not exist at all.

Business intelligence culture may boost business intelligence adoption as well as get in the way of it, for example, create the *rock the boat* syndrome when inconvenient (mostly internal) information is disclosed. Apart from mobilizing human factors, business intelligence culture may serve as an axis of organization-wide information integration, and help the transition from functional to process-oriented mode. The named set of culture components, although far from being exhaustive, reflects a trend towards harmonisation of human factors and alignment of efforts by establishing a flexible set of guidelines for sustainable business intelligence developments.

The important features of BI culture, listed in this chapter, are deemed to be the holding pillars of BI culture. The primary role of BI technologies, systems and digital assets is to provide smooth support in the same way as for information integration (Chap. 5) or management of experience and lessons learned (Chap. 6). The horizontal sharing of intelligence information and insights appears to be the most clearly expressed feature of BI culture, confirmed by empiric research. Horizontal, inter-departmental sharing activities lead to formation of internal intelligence community, bonded by common intelligent assets. Motivational issues remain of prime importance, as BI culture, like other types of culture, can only be built, not implemented as a project.

References

Alavi, M., Kayworth, T. R., & Leidner, D. (2006). An empirical examination of the influence of organizational culture on knowledge management practices. *Journal of Management Information Systems, 22*(3), 191–224.

Appelbaum, S., Calla, R., Desautels, D., & Hasan, L. (2017). The challenges of organizational agility (part 1). *Industrial and Commercial Training, 49*(1), 6–14.

Barton, D., & Court, D. (2012). Making advanced analytics work for you. *Harvard Business Review, 2012*, 79–83.

Boyer, J., Frank, B., Green, B., Harris, T., & Van De Vanter, K. (2010). *Business intelligence strategy. A practical guide for achieving BI excellence.* San Jose, CA: IBM Corporation.

Brijs, B. (2013). *Business analysis for business intelligence.* Boca Raton, FL: CRC Press.

Cameron, K. S., & Quinn, R. E. (2011). *Diagnosing and changing organizational culture: Based on the competing values framework.* Reading, MA: Josey-Bass.

Carlo, J. L., Lyytinen, K., & Boland, R., Jr. (2012). Dialectics of collective minding: Contradictory appropriations of information technology in a high-risk project. *MIS Quarterly, 36*(4), 1081–1108.

Choo, C. W. (2013). Information culture and organizational effectiveness. *International Journal of Information Management, 33*, 775–779.

Choo, C. W. (2015). *The inquiring organization: How organizations acquire knowledge & seek information.* Oxford: Oxford University Press.

Cohen, W. M., & Levinthal, D. A. (1990). Absorptive capacity: A new perspective on learning and innovation. *Administrative Science Quarterly, 35*, 128–152.

Cosic, R., Shanks, G., & Maynard, S. (2012). Towards a business analytics capability maturity model. *Proceedings of the 23rd Australasian conference on information systems.* 3–5 Dec 2012, Geelong, Australia, pp. 1–11.

Cronin, B. (Ed.). (2004). *Annual review of information science and technology* (Vol. 38). Medford, NJ: Information Today.

Davenport, T. (1997). *Information ecology.* New York, NY: Oxford University Press.

Davies, P. (2004). Intelligence culture and intelligence failure in Britain and the United States. *Cambridge Review of International Affairs, 17*(3), 495–520.

Denning, S. (2006). *Seventeen myths of knowledge management.* Retrieved June 15, 2020, from http://www.stevedenning.com/slides/17mythsofkm.pdf.

Elliott, T. (2014). *5 top tips for agile analytics organizations.* Retrieved June 12, 2020, from https://timoelliott.com/blog/2014/09/5-top-tips-for-agile-analytics-organizations.html.

Franklin, M., Halevy, A., & Maier, D. (2005). From databases to dataspaces: A new abstraction for information management. *SIGMOD Record, 34*(4), 27–33.

Gartner Research. (2008). *Gartner reveals nine fatal flaws in business intelligence.* Retrieved July 02, 2020, from https://www.webwire.com/ViewPressRel.asp?aId=77100.

Grublješič, T., & Jaklič, J. (2015). Business intelligence acceptance: The prominence of organizational factors. *Information Systems Management, 32*, 299–315.

Hackatorn, R., & Margolis, T. (2016). Immersive analytics: Building virtual data worlds for collaborative decision suppport. *IEEE VR2016 Workshop on Immersive Analytics*, March 20, 2016.

Hallikainen, P., Merisalo-Rantanen, H., Syvaniemi, A., & Marjanovic, O. (2012). From home-made to strategy-enabling BI: The transformational journey of a retail organization. *Proceedings of the 20th European Conference on Information Systems ECIS 2012,* Barcelona, Spain, June 2012, p. 28.

Hamrah, S. S. (2013). The role of culture in intelligence reform. *Journal of Strategic Security, 6*(5), 160–171.

Harris, J. (2012). *Building an analytical culture for big data.* Retrieved October 03, 2016, from http://data-informed.com/jeanne-harris-how-to-create-analytical-culture/.

Herschel, R. T., & Jones, N. E. (2005). Knowledge management and BI: the importance of integration. *Journal of Knowledge Management, 9*(4), 45–55.

Höglund, L. (1998). A case study of information culture and organizational climates. *Swedish Library Research, 3-4,* 73–86.

Hough, J. (2011). Supporting strategy from the inside. *Journal of the Operational Research Society, 62,* 923–926.

Imhoff, C., & Pettit, R. (2004). The critical shift to flexible business intelligence: What every marketeer wants—and needs—from technology. *An intelligent solutions white paper.* Retrieved June 10, 2020, from http://download.101com.com/pub/tdwi/files/Critical_Shift_to_Flex_BI_Imhoff.pdf.

IT Summit. (2015). Retrieved July 15, 2020, from https://itsummit.lt/bi-day-15.

Johnston, R. (2005). *Analytic culture in the US intelligence community.* Washington, DC: Center for the Study of Intelligence, CIA.

Kiron, D., & Shockley, R. (2011). Creating business value with analytics. *MIT Sloan Management Review,* September 15, 2011.

Lee, M., & Widener, S. (2013). Culture and management control systems in today's high-performing firms. *Management Accounting Quarterly, 14*(2), 11–18.

Marchand, D., & Peppard, J. (2013). Why IT fumbles analytics. *Harvard Business Review,* January–February 2013.

Marchand, D. A., Kettinger, W. J., & Rollins, J. D. (2001). *Information orientation: The link to business performance.* New York, NY: Oxford University Press.

Matzler, K., Strobl, A., & Bailom, F. (2016). Leadership and the wisdom of crowds: how to tap into the collective intelligence of an organization. *Strategy & Leadership, 44*(1), 30–35.

McAfee, A., & Brynjolfsson, E. (2012). Big data: The management revolution. *Harvard Business Review, 2012,* 60–68.

Meredith, R., Remington, S., O'Donnell, P., & Sharma, N. (2012). Organizational transformation through Business Intelligence: Theory, the vendor perspective and a research agenda. *Journal of Decision Systems, 21*(3), 187–201.

Mettler, T., & Winter, R. (2016). Are business users social? A design experiment exploring information sharing in enterprise social systems. *Journal of Information Technology, 31,* 101–114.

Moss, L. T., & Atre, S. (2003). *BI roadmap: The complete project lifecycle for decision support applications.* Boston, MA: Pearson Education.

Mulani, N. (2013). *Take an issues-driven approach to business intelligence.* Retrieved October 06, 2020, from http://www.computerweekly.com/feature/Take-an-issues-driven-approach-to-business-intelligence

O'Reilly, C., & Chatman, J. A. (1996). Culture as social control: Corporations, cults, and commitment. In B. M. Staw & L. Cummings (Eds.), *Research in organizational behavior* (Vol. 18, pp. 287–365). Amsterdam: Elsevier.

Olszak, C. (2016). Toward better understanding and use of BI in organizations. *Information Systems Management, 33*(2), 105–123.

Parr Rud, O. (2009). *Business intelligence success factors: Tools for aligning your business in the global economy.* Hoboken, NJ: John Wiley & Sons.

Popovič, A., Hackney, R., Coelho, P. S., & Jaklic, J. (2012). Towards BI systems success: Effects of maturity and culture on analytical decision making. *Decision Support Systems, 54*(1), 729–739.

Presthus, W. (2014). Breakfast at Tiffany's: The Study of a Successful BI Solution as an Information Infrastructure. *Twenty Second European Conference on Information Systems* (ECIS), Tel Aviv 2014.

Shah, R. (2011). *Shifting the imperative from knowledge management to expertise management.* Retrieved July 02, 2019, from http://www.forbes.com/sites/rawnshah/2011/03/01/shifting-the-imperative-from-knowledge-management-to-expertise-management/.

Skyrius, R., Katin, I., Kazimianec, M., Nemitko, S., Rumšas, G., & Žilinskas, R. (2016). Factors driving business intelligence culture. *Issues in Informing Science and Information Technology, 13*, 171–186.

Skyrius, R., Nemitko, S., & Taločka, G. (2018). The emerging role of business intellkigence culture. *Information Research, 23*, 4. Retrieved December 15, 2019, from http://informationr. net/ir/23-4/paper806.html.

Stangarone, J. (2014). *7 Reasons why BI projects fail*. Retrieved July 02, 2020 from http://www. mrc-productivity.com/blog/2014/05/7-reasons-why-business-intelligence-projects-fail/.

Tyson, K. (2006). *The complete guide to competitive intelligence*. Chicago, IL: Leading Edge Publications.

Villamarin-Garcia, J. M., & Dias Pinzon, B. H. (2017). Key success factors to BI solution implementation. *Journal of Intelligence Studies in Business, 7*(1), 48–69.

Viviers, W., Saayman, A., & Muller, M.-L. (2005). Enhancing a competitive intelligence culture in South Africa. *International Journal of Social Economics, 32*(7), 576–589.

Watkins, M. (2013). What is organizational culture? And why should we care? *Harvard Business Review*, May 15, 2013.

Wells, D. (2008). *Analytic culture—does it matter?* Retrieved November 11, 2016, from http:// www.b-eye-network.com/view/7572.

Yeoh, W., & Coronios, A. (2010). Critical success factors for BI systems. *Journal of Computer Information Systems, Spring, 2010*, 23–32.

Yeoh, W., & Popovic, A. (2015). Extending the understanding of critical success factors for implementing BI systems. *Journal of the Association for Information Science and Technology, 67*(1), 134–147.

Yoon, T. E., Ghosh, B., & Jeong, B.-K. (2014). User acceptance of business intelligence (BI) application: Technology, individual difference, social influence, and situational constraints. *47th Hawaii International Conference on Systems Sciences*, pp. 3758–3766.

Chapter 10
Soft Business Intelligence Factors

10.1 Common Issues for Soft Factors

All three soft factors are discussed together because of their common features: recognized usefulness for advanced informing functions; scarcity and perishability; complicated evaluation and management; dim prospects of automation, at least in the near future; their deficit creates significant overhead. All three factors have been individually researched in many directions—economic, philosophical and critical theoretical perspectives (Crogan & Kinsley, 2012); less so in the terms of their interrelations, and even less so in the current digital realm, let alone the context of BI. There are other factors that have to be prevented from wasting—time, privacy, reputation, and possibly others, but they are the usual features of information quality, as indicated in numerous sources, and their role is rather well researched.

The notion of *attention* is closely related to such important informing problems as information overload and ability to separate useful information from noise. This separation on data level is successfully handled by a string of browsing and retrieval technologies, but more complex information seeking tasks require human *sense* making—Big data technologies that use automatic discovery algorithms are known to produce statistically valid results that make no sense at all. *Trust* in information, or its truthfulness—the important feature of information quality (the opposite is untrue, wrong, incorrect information) has been rather well researched for small elements of information. However, in the current complexity of BI information environment, where information integration from various sources is an everyday need, trust in both sources and product is a feature that has to be protected and preserved. A seemingly important piece of information or information product, which makes very much sense but is not completely trusted, cannot be actionable.

For the above factors, a unifying feature is their limited size or "stock", subject to waste and perishability—all three can be wasted, if relevant measures are not taken for their preservation. Information itself is not subject to waste or perishability, although there are some features that change over time: important information becomes trivial, private information migrates to public etc. As early as 1971, Herbert Simon had famously stated: "a wealth of information creates a poverty of attention" (Simon, 1971). Further in the same source he provided another important statement regarding attention management: "To be an attention conserver for an organization, an information-processing system (abbreviated IPS) must be an information condenser".

Tunkelang (2013) has developed what he called a framework to interact with and benefit from data, consisting of three levels: information, attention, and trust. This framework roughly corresponds to the one discussed here: the *information* component, although understood by Tunkelang as mostly raw data from all possible sources, may be interchanged with *sense* that drives the needs for attention and trust. We should also note here that sense and trust are important features of information quality, whereas attention is a feature of human information behavior.

In the organizational context it is important to understand whether the soft factors are solitary issues, emerge only in groups or between people, or both. Regarding interpersonal issues, or "how many people does it take", the soft issues differ:

- Attention belongs mostly to an individual, although group attention is an item of recent research;
- Trust issues come up in groups and communication—at least two actors are required;
- Sense and sense-making, like attention, is mostly individual; although information sharing supports collective sense-making.

There are (or possibly are) relations between the three that may be important in facilitating advanced informing and removing its obstacles. The following part of this chapter discusses all three soft factors and possible relationships between them.

10.2   Attention

There are numerous definitions of attention in existence, and their semantic analysis goes far beyond the goal of this chapter. A rather adequate example definition by Davenport and Beck (2002) suggests: "Attention is focused mental engagement on a particular item of information. Items come into our awareness, we attend to a particular item, and then we decide whether to act".

10.2.1   Importance for BI

Valliere and Gegenhuber (2013) stated that the information behavior of people who filter the environment for important signals may be used in the organizations by structures that monitor the environment and bring issues to attention—in other words, for the needs of business intelligence. While automatic processing can be used for routine activities, conscious attention is brought to bear only when exceptional cases arise.

Reduced attention may lead to a neglect of important points. As M. Goldhaber put it in 1997, "So a key question arises: Is there something else that flows through cyberspace, something that is scarce and desirable? There is. No one would put anything on the Internet without the hope of obtaining some. It's called attention. And the economy of attention—not information—is the natural economy of cyberspace" (Goldhaber, 1997a).

In 2-cycle diagram of BI and DS (Chap. 3, Fig. 3.4), the move from cycle 1 to cycle 2 happens when something in the cycle 1 attracts attention and deserves a closer scrutiny in cycle 2, with concentration and analytics. This is a very important point in BI, as proper attention management allows to attract attention, and, coupled to sense, detect issues that are important. This ability to detect depends upon many factors: selection of information sources, use of appropriate analytical technology, people competencies, group work and culture, to name a few. Following the 2-cycle model, general attention is used to monitor the environment and satisfy common information needs. Emerging anomalies or other cues attract additional attention and concentration on the issue of interest; special information needs come up requiring approaches that are not for everyday use, and are customized for the issue at hand. Lynch (2005) proposed a model to scale the amount of attention required at different points of time. Focus of attention narrows down with importance of discovered information; more important information gets more space in user's vision field.

A term closely related to attention is concentration which allows to maintain attention for a required time; or it can be said that concentration is sustained attention. Important issues that need concentration mostly relate to complex information needs: they are urgent; carry high-value or high-payoff; are risky or immediately dangerous; or disclose opportunities requiring fast move. For such issues there are no simple answers—the solution or insight is not obvious and will take some effort to clarify, and the result of information integration may be patchy and fragmented.

10.2.2 Important Attention Issues

Information Overload

Information overload is one of the principal sources causing attention waste and perishability. The human ability to process information from the environment experiences problems coping with the deluge of data and information, mostly in digital format. Hallowell (2005) has described attention deficit trait (ADT), brought on by the demands on time and attention that have exploded over the past two decades. As minds fill with noise, the brain gradually loses its capacity to attend fully and thoroughly to anything. According to Hemp (2009), a survey of 2300 Intel employees has shown that people judge nearly one-third of the e-mail messages they receive as unnecessary. A Microsoft study found that the average recovery from interruption by email took 24 minutes. In the same paper, Hemp separated the features of information overload for individuals, where damaged concentration dominates, and for companies that mostly experience lost employee time. Hansman, Kuchar, Clarke, Vakil, and Barhydt (1997) researched attention issues in a rather sensitive area—air traffic management, paying special attention to information and task overload. The paper relates reduction of information overload to automation—aids for filtering, recommendation, decision making. However, automation can be confusing and performance may degrade if underlying automation structure is unclear and does not explicitly signal about its activities.

Most sources, discussing information overload, agree that the growing flow of information creates significant risks by damaging concentration, creating overhead and anxiety. All sources agree that the digital space is the primary source of the overload. A somewhat controversial situation comes up where, in the context of information overload, a "digital silence" from someone, or an inability to find relevant and useful information, lead to anxiety and more loss of concentration.

Distractors, Interruptions, Attention Fragmentation

Another reason for attention waste and perishability is distractors—loud off-topic signals, unnecessary messages and notifications, external interrupts. Kane (2019) presented examples of attention-wasting applications—websites and applications, designed to automatically send mails and messages that are not obvious in their urgency or wide need. Basoglu, Fuller, and Sweeney (2009) researched effects of computer-mediated interruptions on work performance, and detected significant negative indirect influence of frequency of interruptions on decision accuracy through cognitive load for task accuracy. The effect of interruptions on degrading work performance in complex decision tasks has been confirmed by Speier, Valacich, & Vessey (1999). Linda Stone, an executive at Apple and Microsoft, coined the term "continuous partial attention" (Stone, 2009) that appears to be a result of extensive multitasking.

Incorrect Attention Guidance

One more factor in attention waste and perishability is messy interface—incorrectly placed or used attention flags, poorly organized information, no balance between items with different importance. According to Kane (2019), many designs demand more attention than a person can offer—e.g., when asked a direct question, system gives a list of alternative answers instead (asked for driving directions to a certain airport, the system provides alternatives and requires the driver to look at the device screen instead of voice output). To quote H. Simon (Simon, 1996; in Fischer (2012)): "Design representations suitable to the world in which the scarce factor is information may be exactly the wrong ones for situations in which the scarce factor is human attention. In a world in which people suffer from information overload, presented information needs to be relevant to the task at hand and tuned to the background knowledge of the user."

Group Attention Issues

Although attention, similar to decision making, is often attributed to a single-person, there are research sources that discuss group or team attention. Team's attention, just like personal attention, is a scarce resource as well (Birkinshaw, 2015). Fereira, Herskovic, and Antunes (2008) have researched attention management in synchronous electronic brainstorming, and showed that, if backed by a proposed innovative groupware environment, users more smoothly leverage the shifts between individual work and group activities. This leverage resulted in more proposed ideas and less time to formulate and express them. Reviewing a book "The information diet" by C.A. Johnson, Wilson (2012) stated that information overload is not exactly a personal matter: the culture of blame, peer pressure, risk-avoiding behavior and other issues that introduce strain, overhead and degraded performance, can only be dealt with at corporate level.

Metrics

If we admit that attention is limited and therefore scarce, it should possess a scale or measure to evaluate, although at the moment the metrics of attention are hard to define. Theories of human attention all agree that it is limited in capacity (Kane, 2019). There are hints of measurability in everyday use of the term "attention": "pay some attention", "zero attention" or "undivided attention", to mention a few. Many sources point to attention having exchange value: users pay for their service by assigning their attention (Kane, 2019); attention has become a currency for businesses and individuals (Davenport & Beck, 2002); email recipients place a monetary value on their attention when fighting spam (Tunkelang, 2013). There are other points of view: Goldhaber (1997b) had stated that attention can't be bought, and "the attention economy is a zero-sum game. What one person gets, someone else is denied." The statement that the attention cannot be bought is, to author's opinion, somewhat straightforward—while one cannot buy attention in an open market as a commodity, it can be depleted by interruptions and distractors, or expanded by filtering out unnecessary information or sharing the information processing load.

One of the possible indirect approaches to measuring attention, as well as other soft factors in this chapter, is the estimation of expected value of attention (Dasgupta, 2000). Huberman and Wu (2008) presented a "law of surfing" which states that the probability of a user accessing a number of items in a single session markedly degrades with the number of items, and puts a strong constraint on the amount of information that gets explored in a single session. An example is the number of result pages of a Google query that are actually opened—few people go beyond 1–2 pages.

The paper proposes a formal model to combine content rating and assignment of "visual real estate" (usually it is space on a display screen). There are two problems involved: prioritization of content; and finite number of items that a user can attend to (perceive, read, familiarize, consume) in a given time interval. This approach relates to a tiered approach proposed by author in Chap. 4 of this book, which places the most common and often-used items in the first tier of the user environment, and all the remaining items—in the second or more tiers.

The issues of attention metrics emerges in a number of published works; however, a metric of the common type, which can be measured and calculated, hardly exists. Metrics like time or attention marks (clicks, views etc.) are surrogate because of their extensive nature. Time spent on an information item does not necessarily represent the actual amount of attention and consideration assigned to that item. Same can be said about clicks and other actions of selecting an item—they represent only a part of attention dynamics. Attention is a subjective experience; internal distribution and intensity of attention is hard, if not impossible, to track and measure. Many works on attention metrics belong to marketing and e-commerce field. Meanwhile, BI context is not about competing for the attention of a would-be customer, but about coherent guidance of attention to make the most sense of the incoming information, and linking attention to sense.

Regarding attention metrics, it might be useful to look at what atomic actions or events trigger changes in attention. The type of actions or events that decrease attention is more or less clear—distractions and interruptions are the events that cut down attention, or attract attention in unproductive ways. What makes attention increase is the perceived relevance and importance of new information—signal, cue, deviation, anomaly, "black sheep". A simple example of technique for directing attention is exception reports that have been in use for decades. This new information may discover something interesting or promise a prospect of a missing link. Growing attention results in concentration that is expected to get from facts to insight, adding more sense on the way.

10.2.3 Attention Management Tools

The majority of existing attention management tools comes from attempts to manage communication flows emerging from e-mail and other channels. There are tools to manage individual or group attention regarding incoming information flow, as well as to manage outgoing information flow to benefit the attention of recipients of this flow. Several of the most common types of attention management tools are listed below.

Filtering and flow mediation. Likely the most dominant group of proposed attention management tools, aiming at incoming communication messages. An important feature is the user-selected flexibility like selection of filtering criteria.

Automation. Experience and developed rules are used to automate and hide repetitive procedures from the user, thus freeing attention to concentrate on important issues. Such automated functions may include filters, alerts, thresholds, message prioritization.

Informative tagging. Mostly applies to email communication and saves attention by providing informative and unambiguous subject line, allowing for quick user decisions on message importance and urgency. Most email systems contain importance tagging functions; however, tags say nothing about the content.

Signals, cues, alerts. Apply to data stream; a good example is exception reports—based on simple comparison against the preset criteria (e.g., deviation more than X percent). In this case, simplicity is the key—clear data and recognition logic. Recognition of complex issues is handled by more sophisticated techniques like neural networks whose recognition power has been proven in numerous cases. With increasing complexity, however, the reliability of results may raise doubts, especially in the cases of looking for "unknown unknowns".

Managed propagation. As well, mostly applies to email communication, and discourages senders from unnecessary use of "reply to all" function.

Move from push to pull. Directly relates to "reply to all" function, and instead of using it, encourages senders to use notification platforms like Intranet or other sharing environments that do not create false urgency.

Personalization. User-customized informing environments employ automated information management, filtering and personalization systems like Personalization delivery engine (Radhakrishnan, 2010).

Tiered structures. Such structures are part of a personalization approach, and split the information environment of an user into tiers. As presented in Chap. 4, and in author's earlier work (Skyrius, 2008), the first tier would contain the most relevant and often-used elements: key indicators, information sources, functions; while second tier or further tiers would contain more specific elements whose use frequencies are lower—e.g., tools for advanced search and information integration, analytics and modeling, advanced communications and groupwork.

10.3 Sense

The term *sense*, as used in this book and in the current chapter, is interchangeable with the term of *meaning*, and is the expected product of sensemaking, as can be seen from the existing definitions of sensemaking:

- Sensemaking is a process that produces sense (Weick, Sutcliffe, & Obstfeld, 2005);
- Sensemaking can be defined as understanding information or gaining insight into an issue at hand (Pirolli & Card, 2005).

The sensemaking process has been discussed in more detail in the Chap. 5 of this book, and the sense-making issues are woven throughout this book into the materials of each chapter. In this chapter, sense is seen as meaning, or objective reflection of true state of things—a soft static asset that can grow, shrink or be wasted. Together with trust, sense and sensemaking can be seen as the main goal of BI activities. The important experiences from other domains on human factors in intelligence (Heuer, 1999) stress the cognitive challenges and human bias as key factors to consider when relating technology potential to human intelligence capabilities. According to Heuer, the importance of analytical tools and techniques to support higher levels of critical thinking in developing sense in complex issues is obvious. In such situations information is often incomplete, ambiguous, even deliberately distorted. Without the use of analytical techniques for structuring information, challenging assumptions, and exploring alternative interpretations, the development of adequate sense is rather problematic.

Sensemaking is a process that chains a new fact to a structure that's already available. If newly received or discovered information does not clearly relate to previous structures, sense making is difficult. However, in the current wave of data analytics there's a support for confidence in discovered patterns and rules without well-defined articulation of roots of those discoveries. Discussing human sensemaking against the background of contemporary algorithm-based approaches, Madsbjerg (2017) stressed that an old-fashioned process of critical thinking has to regain its weight in organizations to counter-balance the risks of automatic trust in algorithms.

10.3.1   Sense Dynamics

Reception of new information can leave sense unchanged, add sense, or distract some of it by introducing contradiction or confusion. A snippet of information makes sense when it is linked to other information—positioned or benchmarked against some dimensions, relates to known entities—in other words, this new information can be attributed to a place in previous knowledge. Sense should increase when a new snippet of information carries similarity to known context, or un-similarity that is explainable. Similarly, sense may be reduced by incongruent data or information that reduces clarity or contradicts with no obvious explanation. Dervin (1998) has used a term of *sense unmaking* to underline the occurrences of situations that face contradictions or negation of previous sense.

If attention can be wasted by taking away part of it, sense waste and perishability is not caused by disappearance of available information. Instead, it may be initiated by newly received information which proves contradictory or misleading. Environment creates a flow of information that is often controversial—the same messages can contain multiple and sometimes conflicting interpretations. There are numerous cases when additional information brings controversy and confusion, and negates previous efforts in sense building ("it doesn't make sense anymore").

The relation between attention and sense is rather clear at the first look: due to attention we notice something, and then want to recognize and understand. As well, wasted attention leads to sense degradation; and failed sense-making leads to degradation of trust in information, channels or technologies that produced this information. Although subject to waste, sense or meaning is not scarce in the way attention is.

The substantial available research on sense suggests that, despite its static nature at a given moment of time, it is ever-changing and never final. "This is clear at last"; "it is no longer true"; "it depends (upon context)"—these are the statements from everyday activities that justify sense dynamics.

One area where sense handling has made significant advances automating sense decoding is text mining and analytics. Seth Grimes (2007), describing a history of text analytics, provided a point for evaluating the advances in the area:

> Recently, a new, different style of text analysis has grown in prominence: the use of familiar business intelligence interfaces and techniques to gain a broad understanding of trends without excessive concern for individual cases—for those needles in haystacks. And recently, feature-extraction capabilities have moved beyond entities such as names and e-mail addresses to events and sentiments. When software can parse "the service was lousy" and initiate an appropriate action, and when it can start to comprehend the non-literal human meaning of a phrase such as "Yeah, that idea is a real winner," then we've made real progress in achieving H.P. Luhn's 1958 goals for business intelligence systems....

Repetitive reception of the same information does not add sense, but may add trust by triangulation—a procedure to test the reliability of information by comparing same information from different sources, or produced by different methods. A methodology of triangulation, ACH (Analysis of Competing Hypotheses) has been presented in (Moore, 2011).

10.3.2 Sense Metrics

The attempts to measure sense or meaning have been discussed in Chap. 5, and several examples of candidate units have been presented. These units range from simplest (sign, term, note, comment) to complex (connections, coincidences, curiosities, contradictions), and all of them are surrogate units regardless of complexity—they represent a static view, borrowed from material assets, with intent to isolate an element of sense without relating it to other elements, movements or mutations over time.

If a generic unit is, for example, a statement, several statements of about the same size and look may significantly differ in how they affect the available sense on situation:

– Relevance—relation to the meaning of other statements centered around a current concern, that adds new facts or angles;
– Importance—what does an important received statement do, and why is it considered important? Does it change the general understanding of the situation at hand? Does it reduce confusion significantly or remove obstacles for an important decision?

– Clarity—a feature of the statement (clear or unclear) or a meaning of situation (new statement added clarity to the existing understanding).

While discussing measuring meaning in sociological studies, Mohr (1998) pointed out that attempts to model meanings or subject it to formal analysis involve a gross simplification of the material. Despite the risk of this simplification, a number of interesting approaches to introduce measures and quantitative methods has been proposed in this area. As an example, Moore (2011: 135–137) describes a model to conceptualize measurement of rigor in sensemaking, proposed by Zelik, Patterson, and Woods (2007). The proposed model joins eight elements and three levels for each element—high, medium and low. The elements of the model are:

- Hypothesis Exploration—defines how broad was the evaluation of alternatives;
- Information Search—evaluates the depth and breadth of the search process in terms of number and proximity of data sources;
- Information Validation—details the levels of information cross-validation;
- Stance Analysis—evaluation of source reliability and bias;
- Sensitivity Analysis—the extent to which the analyst understands the assumptions and limitations of analysis;
- Specialist Collaboration—the degree to which the opinions of domain experts have been incorporated;
- Information Synthesis—the degree of how far did the information integration reach;
- Explanation Critique—the level of incorporation of perspectives of different experts in examining the primary hypotheses.

The model does not measure sense or meaning directly, nor does it suggest any well-defined metrics for rigor of intelligence sensemaking. However, it does provide an attempt to cover the key directions of sensemaking activities, and gives rather clear idea of what are the opposites on the dimension of each element, thus providing a framework to evaluate quality of sensemaking process. Evidently, the approach is intended for complex or non-routine informing situations, featuring the attributes of complex information needs: multiple information sources, multiple procedures of handling source information, multiple participants of the process.

10.4 Trust

Trust in informing activities is profoundly important as an important part of product value. Like attention and sense, it is perishable, though not as evident in scarcity as attention. The issues of trust in the fields of psychology and organizational behavior are rather well-researched, which cannot be said about trust in the field of informing, information systems and business intelligence in particular. In the contemporary technology-supported mesh of communications and collaboration, the need for trust has become a basic ingredient for inter-organizational success (Asleigh &

Nandhakumar, 2005). Trust is a critical factor of likelihood that entities will transact and interact in a digital environment (Pranata, Skinner, & Atauda, 2012). Positive trust is a catalyst for human cooperation, and a very important feature of BI product quality (Jøsang, Keser, & Dimitrakos, 2005). Trust is central to all business transactions, and yet economists rarely discuss the notion of trust (Dasgupta, 2000).

Scheps (2008) has provided a rather blunt description of trust in BI activities: "Nothing's more frustrating in the world of business intelligence than a development team toiling for months to produce a report that an executive looks at and, within 30 s, dismisses it by saying, "Those numbers aren't correct.""

10.4.1   Definitions

The general definitions of trust are abundant, and focus mostly on trust between people:

- Trust is "the willingness of a party to be vulnerable to the actions of another party based on the expectation that the other will perform a particular action important to the trustor, irrespective of the ability to monitor or control that other party" (Mayer, Davis, & Schoorman, 1995; in Dennis, Papagiannidis, Eleftherios, & Bourlakis, 2016).
- Trust is a condition for effective cooperation among individuals, groups and organizations (Asleigh & Nandhakumar, 2005).

Trust is never singular, and is exercised in groups and communication, where there are no less than 2 actors: trustors—entities giving trust, or those who trust; and trustees—entities receiving trust, or those who are trusted (Söllner, Hoffmann, & Leimeister, 2016). The trustor uses elements of trust (context, reputation, belief, time) to estimate the level of trust in trusted entity. The important characteristics of trust, according to (Pranata, Skinner, & Atauda, 2012), are:

- Trust is **dynamic** and applies only for a given time period;
- Trust is **context-dependent** and relates to a given domain;
- There is **no full trust**—never 100%;
- Trust is **transitive** within given context—A trusts B, and B trusts C, so A trusts C; hard to quantify;
- Trust is **asymmetric** in nature—if A trusts B, B not necessarily trusts A.

It can be noted that some of the above statements are subject to discussion. E.g., regarding full trust, the authors assume that trust is never complete, and allow estimations below 100%. There are opposing opinions—Dasgupta (2000) states that trust can be seen as an absolute medium—trust is there, or there is no trust.

Chasin, Riehle, and Rosemann (2019) have pointed out several types of trust relations:

- Interpersonal trust;
- Trust between individuals and organizations;
- Trust in technology;
- Intra-organizational trust;
- Inter-organizational trust;
- Trust in institutions.

We can distinguish between interpersonal trust (between people and groups of people), and trust in informing activities, including trust in information technology, systems, and the information these systems provide as a result. The latter also covers the issues of trust in business intelligence, where the often-encountered complexity of informing places trust among the most important components of product quality. There are numerous examples of trust in digital space: trust infrastructures on the Net—e.g., credit card systems, e-banking, user comments on TripAdvisor, and many others.

In informing activities trust is exercised by user or users, whose trust is divided among the elements of informing process: source data and its provenance, delivery channels, processing methods and systems design, other contributors, security levels and so on. In this context, waste and perishability of trust may be caused by all possible reasons leading to wrong information—human and technical, intended and unintended. Trust is part of information culture, according to Abrahamson and Goodman-Delahunty (2013). The provided composite definition of information culture ties, albeit indirectly, trust and information sharing.

The importance of trust in BI environment is boosted by the sensitivity of BI product, and by often-political nature of its use. BI insights may introduce surprises and discover inconvenient truths that may be dangerous to groups within an organization. People, who find the BI results embarrassing to them, most likely will challenge these results, and any errors will introduce doubts about the entire BI function. Especially in the beginning of BI activities, its products have to be easily checkable, traceable and have clear provenance to develop trust over time. The simpler logic of a system, the faster would be development of trust—it is hard to trust a system for producing important information with complicated and opaque logic.

The above risks of internal disclosure can be mediated with open information culture that would facilitate internal trust. Accenture (Long, Roark, & Theofilou, 2018) has developed a Competitive Agility Index that comprises in equal parts growth; profitability; sustainability and trust. The growing role of transparency can be projected into internal transparency, based on flexible informing and trust as a factor of shared value. Trust incidents result in loss of trust from the perspective of stakeholders. The Index defines six groups of stakeholders: employees, customers, suppliers, media, analysts and investors. Regarding employees as an internal group (which is a creator and user of BI at the same time), ranked reputation of a company is among top-three motivators to work for this company. The study states that a drop in trust impacts financial results, and presents evidence in several cases.

10.4.2   Trust in Elements of Informing Process

Source Data

One of the weakest links in informing chain is the quality and reliability of source data. Terms like believability (Stock & Winter, 2011) and provenance (Ram & Liu, 2008) are used in the same context. According to Stock and Winter, believability evaluates to which degree the information is trustworthy; high believability leads to increased acceptance and use of BI systems, and contributes to better decision making. Discussing provenance, Ram and Liu (2008) stated: "... Identifying the source of data enables an analyst to check the origins of suspect or anomalous data to verify the reality of the sources or even repair the source data. In business intelligence and data warehousing applications, provenance is used to trace a view data item back to the source from which it was generated. ... Tracing the origins of a view data item enables users to detect the source of bad data".

Data quality and its impact on overall trust in BI goes well beyond the scope of this book. Referring to the earlier research by the author (Skyrius & Winer, 2000) on the use of information sources for decision making, a fraction of its results is presented in Table 10.1 below.

Although the sample size both for respondents from USA and Lithuania has been modest, we can see a rather similar distribution for both samples. The most often-used information sources are the closest and the most reliable, as they come from internal information environment, have been used and tested before, and an adequate level of trust has been developed for these sources.

The current rise of Big Data analytics has raised huge expectations of discoveries in large data volumes, and a number of practical issues has emerged together with interest. One of the substantial issues is the trust in results of analysis. SAP evangelist Timo Elliott (2018) has presented several key concerns regarding trust in Big Data analytics:

- Data completeness and accuracy. Many data collections contain significant uncertainty and imprecision, especially with unstructured data like user comments.
- Data credibility. External data may contain biases or false values, and data credibility is hard to evaluate.

Table 10.1 Distribution of decision information sources (Skyrius & Winer, 2000)

Information sources	Number of cases	
	In the USA	In Lithuania
Own resources	15	28
Personal contacts	13	27
External computerized sources	10	18
Mass media	4	17
Consultants	8	13
Other	–	6

- Data consistency. While traditional database systems use strict constraints to maintain consistency, some big data systems relax these constraints.
- Data processing and algorithms. Low-level captured data requires intense use of processing algorithms, whose logic has more potential for bias and incorrect conclusions.
- Data validity. Even for accurate source data, it may be inappropriate for a specific business question. Well-pointed formulation of business questions to produce reliable results is one more area where sense and sensemaking meet trust.

Delivery Channels

It is obvious that trust changes along the dimension of external/internal information sources; and although there are examples of unreliable data delivered from internal sources via internal channels, most of the cases with data of unknown reliability are attributed to external channels. To estimate the trustworthiness of (mostly) external data, methods like data triangulation are applied whenever possible. The benefits of triangulation are listed in (Triangulation, n.d.):

(a) Data from different sources can enrich each other by explaining different aspects from different perspectives;
(b) Different data may contradict each other; the composite information is deemed unreliable until clarification;
(c) If data from different and independent sources is consistent, it confirms the accuracy;
(d) Data from one source can explain inconsistent data from another source (see item (a)).

The same source presents a set of triangulation directions:

- Independent data sources;
- Different investigators;
- Different methods of data collection;
- Different conceptual frameworks—essentially, different aspects on the same data, or questions from different domains;
- Time and locations.

Söllner, Hoffmann, and Leimeister (2016) elaborate on the idea of a network of trust, proposed by Muir (1994), and discuss interplay between multiple targets of trust, projecting it into the technology acceptance framework (TAM). The presented model relates TAM and network of trust, including trust in the Internet, trust in the Information system, trust in the community of Internet users, and trust in the Provider. Their results indicate that people tend to trust other actors of a trusted environment (not the technology of a trusted environment) more than actors of a non-trusted environment.

10.4.3 Processing Methods and Systems Design

Information systems should be a reliable source of trusted information because of their deterministic nature, and once technology is trusted, it will maintain this trust by precisely performing what it is expected to do. However, this is far from always being the case. With organizations opening up their information boundaries, information is scattered over multiple sites and locations with different provenance and standards. Such stretched systems are counterintuitive, and trust in a technology or a system can be hurt by inadequacy to cope with changing conditions. Discussing trust in information systems, DeLone and McLean (2016) note the changes in the scope of information systems and its impact on trust:

> It is important to note that, as the foci of our systems have moved outside of our organizational boundaries to customers, citizens, patients, contractors etc., the element of trust has become a more significant "System Quality" and "Information Quality" attribute. Do customers trust the security of information systems? Do citizens trust the information provided by government systems to be accurate and unbiased? Although traditional measures of systems and information accuracy and reliability may suffice here, we should explore the possibility of new measures of system and information trust.

Traditional information systems with well-defined logic may employ justification (Losee, 2014) as the consistency measured between a statement and some related benchmark statements. For information systems having intense external information flows (BI systems fall into this category), product complexity and often unique nature may significantly limit the availability of benchmark information.

Thielsch, Meessen, and Hertel (2018) focused on the trust relation between employees and internal IS, as opposed to dominating research focus on e-commerce and Web services. The paper investigated the dependence of trust or distrust in the information system from:

- User technology competence, trust in technology, regular frequency of use;
- System quality (reliability, controls, ease of use, response time);
- Information quality (volume, clarity, credibility, relevance, security, informative content);
- Service quality (support);
- Context (participation, transparency, error communication, perceived organizational support, obligation to use); and
- Persons involved (ability, attitude, accountability),

The desired outcomes are: well-being, stress, performance, post-situational frequency of use; post-situational satisfaction of use; situational trust or distrust. The proposed model is inwards-oriented; authors suggest that predictors of trust in Web applications such as e-commerce are different from those for internal IS intended for everyday work tasks. However, BI, due to its overarching nature that covers internal and external orientation, relates to both types of activities. Consequently, trust in BI should experience the influence of factors from both approaches.

10.4.4 Contributors

Social dimension of trust is present when we have to rely on colleagues and other human sources to serve as information filters. Trust may be considered as being a part of relationship capital. Asleigh and Nandhakumar (2005) have listed important trust constructs within groups:

- Within team: honesty, understanding, respect, quality of interaction and confidence, pro-activity/reliability/communication, teamwork/commitment, performance/ability, expectancy;
- Between teams—quality of interaction, understanding, teamwork, honesty and confidence, communication and reliability, ability, commitment/respect/expectancy/performance, pro-activity.

The above constructs largely overlay with cultural issues, discussed in Chap. 9, and may serve as guidelines for possible research on relation between information culture, BI culture and trust, and its impact on organization performance in general.

10.4.5 Metrics

Dasgupta (2000) suggests that trust is a commodity; it has no measurable units, but its value can be measured in terms of what benefits it brings to us. Jøsang et al. (2005) discussed trust as a factor of value creation—lack of trust makes us waste time and resources on protecting ourselves against possible harm and creates significant overhead. The activities of compliance, audit, testing in maintaining and developing trust create overhead that can be seen as cost of trust, or a value assigned to trust.

There are proposals to use subjective probabilities of trusted behavior, as they reflect the expectancy held by an individual that the promise of another individual or group can be relied on (Dasgupta, 2000). In producing a composite insight, where information from different sources of varied provenance is mixed into a joint meaning, it would be convenient to have a composite rating of trust in the produced insight, most likely evaluated by probabilities and weights.

10.5 Relations Between Soft Factors

Being important components of informing process, the three soft factors are expected to be related, and we can assume that better understanding of these relations would help to gain better understanding of human issues in informing. Simple attempts to produce relations may look like the two following examples (Table 10.2):

Table 10.2 Example statements tying together attention, sense, and trust (source: author)

Statement	Attention	Sense	Trust
The mission of BI is ...	not to miss ...	... important issues ...	... that can be relied on in important decisions
The system is difficult to use—	... it's hard to concentrate attention	... products or results make little sense	... and it is not clear whether they can be trusted

In D. Tunkelang's scheme, relating information, attention and trust (Tunkelang, 2013), information comes first as a source material for all rational activities. It is followed by attention that acts as a filter to balance our information consumption, and its space or volume is limited. Trust is placed on the top of Tunkelang's pyramid to certify the safety and integrity of consumed information. Two important dimensions of trust are proposed—authority and sincerity. Authority is the source's ability to deliver objective truth. Sincerity is the source's desire to deliver subjective truth. The two mechanisms for building trust are history and networks. For history, we assume that past trustworthiness is predictive of the future trend, and heavily penalize even a single breach of trust. Networks regulate (grow or reduce) trust among participants—people, groups and systems.

The pairwise binary relations between the soft factors have not been researched empirically in any way, so the below statements are speculative, and their goal is to support initial understanding of relations between the factors. However, the most meaningful ones can be tested empirically in the future.

Attention-sense. This relation is rather natural and straightforward: if an object has attracted attention, the attention subject attempts to make sense of it. An example would be a detection of a market opportunity.

Sense-attention. Such are the cases when attention is attracted by familiar cues, signs or other snippets of information whose meaning is known beforehand.

Attention-trust. This relation should be rather weak, unless attention is given to find, for example, a trusted and tested source of information. In addition, information deluge complicates trust—to produce trusted insights, more procedures of selection, filtering, comparison, justification are needed.

Trust-attention. Such case is quite common in decision support in a problem analysis stage, where the known trusted information sources are the ones to begin with.

Sense-trust. A rather common relation where sense does not necessarily lead to trust. For example, a useful-looking collection of data on prospective customers may prove to be flawed and not trustworthy at all. An interesting instance of this relation is the relation between relevance (sense) and rigor (trust), discussed in the next chapter regarding research issues in the field.

Trust-sense: lost trust reduces sense. If some part of information on a problem seemed reliable and fitting, yet further inquiries have proven it erratic or false (loss of trust), it does not make sense anymore—the overall sense is reduced. It is potentially interesting that the confirmed trust does not add any new sense—we do not learn anything new, but trusted information becomes more valuable.

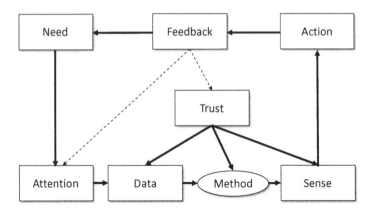

Fig. 10.1 Model of interrelations between soft factors (source: author)

Summing up, one of the possible models of interrelations between the soft factors in informing process is presented in the Fig. 10.1. Models of similar nature have been proposed by T. Wilson, C. Kuhlthau, C. W. Choo.

Here the cycle starts with an information need that activates attention. Attention leads to data, beginning with the most likely and trusted places. Method contains the processing logic and procedures that produce sense, and ultimately—insight. Sense leads to action, although not in all cases—sense may just improve awareness, leaving action for the future. Action tests and justifies sense, providing feedback for directing attention and applying trust in the future.

Regarding how all soft factors apply to BI systems and their product, if valuable sense that justifies investment in BI is hard to produce, trust and attention for BI information degrade.

A case of rearranging BI activities to maximize sensemaking and trust has been described in (Mayer & Schaper, 2010):

From information to intelligence

One global corporation decided that the best way to tackle these problems was a wholly redesigned IT blueprint to support top management. The company, a multibillion-dollar global logistics organization prized for its ability to transport goods from one corner of the globe to another, was having a tough time getting its internal executive information system in order. While it could track cargo along any given point of its delivery network, it had little visibility into its own data streams. Years of rapid growth and decentralized, somewhat laissez-faire information management had created an untidy patchwork of reporting processes across its divisions. Management lacked a single viewpoint into the company's core performance data and, as a result, couldn't know for sure which products made money.

Knowing that something had to be done, the CIO formed a task force, with members from both the business side and IT, which quickly found that relations between them were in some ways dysfunctional. Executives from headquarters, the business units, and the divisional and central IT functions all documented performance in their own way, tapping into different data sources to tally their results. These figures were rolled up into a series of group reports, but variances in the underlying data and the lack of a uniform taxonomy made it difficult for managers to know which set of numbers to trust. The management interface, designed to present key performance data, was jammed with so many different and, in some cases, conflicting KPIs as to be largely unusable. A non-user-friendly front end compounded the problem. Executives therefore asked for a new executive information system to gauge the company's performance at varying levels of detail (exhibit). This new corporate navigator would have to incorporate major improvements in design and functionality.

The CIO at a global company needed to take an end-to-end view in designing his company's business intelligence system. At the front end, reports needed to cascade in a logical hierarchy. The top managers' interface required an intuitive design offering a wide range of analyses. To support this robust interface, the underlying information structure would have to be reconceived. Data were standardized and information pathways reworked for the most important key-performance indicators (KPIs) needed to manage the company. The corporate navigator is its framework for designing a next-generation executive information system . . .

. . . The entry screen, which gives executives a graphical top-line summary of the most important KPIs (Layer 1), includes comment fields to highlight findings or explain deviations. Using the one-page reporting format, Layer 2 offers an at-a-glance assessment of corporate performance in four core areas: financial accounting, management accounting, compliance, and program management. With a few mouse clicks, anyone interested in the underlying details can access pre-defined analyses, such as contribution margins and P&L calculations, by product or business unit. A hierarchical reporting structure connects these views, allowing users to toggle through different levels of performance data.

In the above case, the redesign of business intelligence system has taken into account attention support issues by providing a layered design of the information space and placing there the most important KPIs. Clear information hierarchy and navigation assist sensemaking, while standardized requirements for information structure add trust in the system.

10.6 Possible Research Directions

All three soft factors, because of their vague nature, are quite complicated regarding their research by rigorous methods.

The majority of current research on attention management concentrates on issues like information overload and attention fragmentation, influence of interruptions on work performance. The volume of research done on attention issues in BI is, however, limited. The research of attention issues in detection of important trends in the changing environment does not seem to lose its relevance. The current wave of interest in Big Data analytics and renewed interest in AI potential raises interesting issues about balancing human attention with automated discovery potential. Here, the complexity of algorithmic approaches requires to address problems like false recognition or neglect of important issues. Another interesting area is group attention and attention issues inside BI community. Sense management encompasses modestly researched areas like information integration, context management, combining them with NLP (natural language processing) and other advanced information management techniques. While attention as a scarce resource has a perceivable limit, and therefore its waste is easier to detect, same cannot be said about sense— its growth or waste are much more difficult to comprehend, and have been much less researched. Trust issues are well researched, but less so in the context of BI, and even less so in relation with other soft factors discussed here.

One more possible assumption for further research may be related to sharing, groups and communities whose influence might improve the levels of all soft factors:

- Better distribution of attention because of perceived importance of information circulating in the group; this information might have been selected by other group members;
- More sense or better sense—because of shared sense-making of information circulating in the group; this information might have been defined, evaluated and integrated by other group members;
- More trust because, according to Tunkelang (2013), networks propagate trust among people and other information sources; enable trust to operate at scale, but are vulnerable to deception; better trust because of perceived reliability/credibility of information circulating in the group; this information might have been tested/ verified by other group members.

References

Abrahamson, D. E., & Goodman-Delahunty, J. (2013). The impact of organizational information culture on information use outcomes in policing: An exploratory study. *Information Research, 18*, 4. Retrieved March 12, 2020, from http://informationr.net/ir/18-4/paper598.html#. X37Ljmj7Tcs.

Asleigh, M. J., & Nandhakumar, J. (2005). Trust and technologies: Implications for organizational work practices. *Decision Support Systems, 43*, 607–617.

Basoglu, K. A., Fuller, M. A., & Sweeney, J. T. (2009). Investigating the effects of computer mediated interruptions: An analysis of task characteristics and interruption frequency on financial performance. *International Journal of Accounting Information Systems, 10*, 177–189.

Birkinshaw, J. (2015). Manage your team's attention. *Harvard Business Review, 29*, 2015.

Chasin, F., Riehle, D. M., & Rosemann, M. (2019). Trust management: An information systems perspective. *Proceedings of Twenty-Seventh Conference on Information Systems (ECIS2019)*, Stockholm-Uppsala, Sweden.

Crogan, P., & Kinsley, S. (2012). Paying attention: Towards a critique of the attention economy. *Culture Machine, 13*, 1–29.

Dasgupta, P. (2000). Trust as commodity. In D. Gambetta (Ed.), *Trust: Making and breaking cooperative relations, electronic edition* (Vol. Chapter 4, pp. 49–72). Oxford: Department of Sociology, University of Oxford.

Davenport, T. H., & Beck, J. C. (2002). *The attention economy: Understanding the new currency of business*. Boston, MA: Harvard Business Press.

DeLone, W. H., & McLean, E. R. (2016). Information systems success measurement. *Foundations and Trends in Information Systems, 2*(1), 1–116.

Dennis, C., Papagiannidis, S., Eleftherios, A., & Bourlakis, M. (2016). The role of brand attachment strength in higher education. *Journal of Business Research, 69*, 3049–3057.

Dervin, B. (1998). Sense-making theory and practice: An overview of user interests in knowledge seeking and use. *Journal of Knowledge Management, 2*(2), 36–46.

Elliott, T. (2018). *How trustworthy is big data?* Retrieved July 31, 2020, from https://www. brinknews.com/how-trustworthy-is-big-data/.

Fereira, A., Herskovic, V., & Antunes, P. (2008). Attention-based management of information flows in synchronous electronic brainstorming. In R. O. Briggs et al. (Eds.), *Groupware: Design, Implementation, and Use, 14th International Workshop* (pp. 1–16). Omaha, NE, USA: CRIWG.

Fischer, G. (2012). Context-aware systems—the 'right' information, at the 'right' time, in the 'right' place, in the 'right' way, to the 'right' person. *Proceedings of AVI'12—International Working Conference on Advanced Visual Interfaces*, May 21–25, 2012, Capri, Italy, pp. 297-294.

Goldhaber, M. (1997a). The attention economy and the Net. *Conference on Economics of Digital Information*, Kennedy School of Government, Harvard University, Cambridge, MA, January 23–26, 1997. Retrieved July 12, 2020, from https://firstmonday.org/article/view/519/440

Goldhaber, M. (1997b). *Attention shoppers!* Retrieved July 11, 2020, from https://www.wired.com/1997/12/es-attention/.

Grimes, S. (2007). A brief history of text analytics. *Business Intelligence Network, 30*, 2007.

Hallowell, E. (2005). Overloaded circuits: Why smart people underperform. *Harvard Business Review, 83*(1), 54–62.

Hansman, R. J., Kuchar, J. K., Clarke, J.-P., Vakil, S., & Barhydt, R. (1997). Integrated human centered systems approach to the development of advanced air traffic management systems. *16th DASC. AIAA/IEEE Digital Avionics Systems Conference. Reflections to the Future. Proceedings*, Irvine, CA, USA, 1997, pp. 6.3-1, doi: https://doi.org/10.1109/DASC.1997.636183.

Hemp, P. (2009). Death by information overload. *Harvard Business Review, 2009*, 83–89.

Heuer, R. J. (1999). *Psychology of intelligence analysis. Center for the Study of Intelligence*. Langley, McLean: CIA.

Huberman, B. A., & Wu, F. (2008). The economics of attention: Maximizing user value in information-rich environments. *Advances in Complex Systems, 11*(4), 487–496.

Jøsang, A., Keser, C., & Dimitrakos, T. (2005). Can we manage trust? *Proceedings of the 3rd International Conference on Trust Management (iTrust)*, Paris, May 2005.

Kane, L. (2019). *The attention economy*. Retrieved July 14, 2020, from https://www.nngroup.com/articles/attention-economy/.

Long, J., Roark, Ch., & Theofilou, B. (2018). *The bottom line on trust. Accenture*. Retrieved July 15, 2020, from https://www.accenture.com/_acnmedia/thought-leadership-assets/pdf/accenture-competitive-agility-index.pdf.

Losee, R. (2014). Information and knowledge: Combining justification, truth, and belief. *Informing Science, 17*, 75–93.

Lynch, K. (2005). *Attention management*. Retrieved July 12, 2020, from http://www.klynch.com/archives/000080.html.

Madsbjerg, C. (2017). *Sensemaking: The power of the humanities in the age of the algorithm*. New York, NY: Hachette Books.

Mayer, J., & Schaper, M. (2010). *Data to dollars: Supporting top management with next generation executive information systems. McKinsey Digital*, January 1, 2010. Retrieved July 31, 2020, from https://www.mckinsey.com/business-functions/mckinsey-digital/our-insights/data-to-dollars-supporting-top-management-with-next-generation-executive-information-systems.

Mayer, R. C., Davis, J. H., & Schoorman, F. D. (1995). An integrative model of organisational trust. *Academy of Management Review, 20*(3), 709–734.

Mohr, J. W. (1998). Measuring meaning structures. *Annual Review of Sociology, 24*, 345–370.

Moore, D. T. (2011). *Sensemaking. A structure for intelligence revolution*. Washington, DC: National Defense Intelligence College.

Muir, B. M. (1994). Trust in automation: Part I. Theoretical issues in the study of trust and human intervention in automated systems. *Ergonomics, 37*(11), 1905–1922.

Pirolli, P., & Card, S. (2005). The sensemaking process and leverage points for analyst technology as identified through cognitive task analysis. *Proceedings of International Conference on Intelligence Analysis*, May 2005, McLean, VA.

Pranata, I., Skinner, G., & Atauda, R. (2012). A holistic review on trust and reputation management systems for digital environments. *International Journal of Computer and Information Technology, 1*(1), 44–53.

Radhakrishnan, R. (2010). *Are you paying attention, or is it the other way round?* Retrieved November 27, 2013, from https://www.infosysblogs.com/eim/2010/02/are_you_paying_atten tion_or_is.htm.

Ram, S., & Liu, J. (2008). A semiotics framework for analyzing data provenance research. *Journal of Computing Science and Engineering, 2*(3), 221–248.

Scheps, S. (2008). *Business intelligence for dummies.* Hoboken, NJ: Wiley Publishing.

Simon, H. (1971). *Designing organizations for an information-rich world.* Retrieved July 31, 2020, from *digitalcollections.library.cmu.edu.*

Simon, H. (1996). *The sciences of the artificial* (3rd ed.). Cambridge, MA: The MIT Press.

Söllner, M., Hoffmann, A., & Leimeister, J. M. (2016). Why different trust relationships matter for information systems users. *European Journal of Information Systems, 25*(3), 274–287. https://doi.org/10.1057/ejis.2015.17.

Skyrius, R. (2008). The current state of decision support in Lithuanian business. *Information Research, 13*(2), paper 345. Retrieved March 2, 2020, from http://InformationR.net/ir/13-2/paper345.html.

Skyrius, R., & Winer, C. (2000). IT and management decision support in two different economies: A comparative study. In *Proceedings of IRMA2000: Challenges of information technology management in the 21st century* (pp. 714–716). Anchorage, Alaska, USA.

Speier, C., Valacich, J. S., & Vessey, I. (1999). The influence of task interruption on individual decision making: An information overload perspective. *Decision Sciences, 30*(2), 337–360.

Stock, D., & Winter, R. (2011). The value of business metadata: Structuring the benefits in a business intelligence context. In A. D'Atri, M. Ferrara, J. F. George, & P. Spagnoletti (Eds.), *Information technology and innovation trends in organizations* (pp. 133–141). Heidelberg: Springer.

Stone, L. (2009). Beyond simple multi-tasking: Continuous partial attention. Retrieved on March 12, 2020, from https://lindastone.net/2009/11/30/beyond-simple-multi-tasking-continuouspartial-attention/.

Thielsch, M. T., Meessen, S. M., & Hertel, G. (2018). Trust and distrust in information systems at the workplace. *PeerJ, 6,* e4583. https://doi.org/10.7717/peerj.5483.

Triangulation (n.d.). *INTRAC for civil society.* Retrieved July 31, 2020, from https://www.intrac.org/wpcms/wp-content/uploads/2017/01/Triangulation.pdf.

Tunkelang, D. (2013). *Information, attention, and trust. Interactive*: Retrieved July 31, 2020, from https://www.linkedin.com/pulse/20130102083704-50510-information-attention-and-trust/.

Valliere, D., & Gegenhuber, T. (2013). Deliberative attention management. *International Journal of Productivity and Performance Management, 62*(2), 130–155.

Weick, K. E., Sutcliffe, K. M., & Obstfeld, D. (2005). Organizing and the process of sensemaking. *Organization Science, 16*(4), 409–421.

Wilson, T. D. (2012). Review of: Johnson C.A. (2012) The information diet: A case for conscious consumption. *Information Research, 17*(1), R432. Retrieved July 7, 2020, from http://informationr.net/ir/reviews/revs432.html.

Zelik, D., Patterson, E. S., & Woods, D. D. (2007). Understanding rigor in information analysis. *Proceedings of the Eighth International NDM Conference,* Pacific Grove, CA, June 2007.

Chapter 11
Encompassing BI: Education and Research Issues

11.1 The Encompassing Business Intelligence

This book contains an approach to BI as an organization-wide informing environment for advanced informing, comprising intelligence, analytics and decision support. This approach makes an attempt to unite the value-creating features of previous waves of advanced informing with current developments. There are aspects that stand the test of time, and they are permanently important:

– User information needs;
– Source data quality;
– Understanding of IT capabilities;
– Understanding of the role of human factors;
– Seeing informing activities as an ever-changing process—no system is ever final;
– In a changing world, BI is doomed to be permanently immature.

Terms like "holistic", "360 degrees" and other have started being applied to BI relatively recently, yet they are reflecting a familiar need to learn the entirety of an object or problem of interest. "Encompassing" is another term that may be used interchangeably here; it means the expected wide coverage of key activity aspects by BI. This coverage should include:

– **Internal** issues to be aware of current situation in organization at every level;
– **External** issues at micro-level (customers, competitors, suppliers, institutions) and macro-level (raw material prices, currency rates, inflation), also technology landscape and innovations;
– **Stakeholder interests** and expectations;
– Other sources of important changes;
– A holistic **response plan** based on organizational agility traits and culture.

R. Skyrius, *Business Intelligence*, Progress in IS,
https://doi.org/10.1007/978-3-030-67032-0_11

The coverage is never complete, as limitations of time, money, technology and other resources come up against the need to "leave no stone unturned". The intended coverage is more like a set of adequately complete information for organizational awareness. A portfolio principle often emerges in terms of structuring resources in a best possible, or good enough way against principal goals—better informing, or stronger organization.

A good informing system, paraphrasing March and Hevner (2007), should make the executives questions their assumptions about the environment, and ask the right questions. BI culture, discussed in Chap. 9 of this book, is expected to provide substantial support for encompassing coverage by openness and readiness to face informing challenges. A related concept along the same lines is Peter Senge's framework for the learning organization (Parr Rud, 2009), supporting encompassing intelligence with five disciplines:

A. Systems Thinking—see the system in entirety, not its parts in isolation.
B. Personal Mastery—a special level of proficiency or self-actualization.
C. Mental Models—deeply ingrained assumptions, generalizations or images on how we understand the world.
D. Building Shared Vision—shared goals and values.
E. Team Learning—teams have to connect and share through dialogue while suspending assumptions and learning to trust each other.

The above disciplines conform rather well to BI understanding presented in this book—hybrid, augmented, or assigned whatever other label, BI is supposed to boost human intelligence capabilities by providing a best possible join between potentials of people and technology, and eventually developing or approaching required coverage. This logic has been followed in the book, and is briefly reviewed below.

11.2 Review of the Book Logic

Chapter 2 is dedicated to the definitions and boundaries of BI as an advanced informing environment, whose primary goal is to satisfy complex information needs—clarify complicated situations and answer tough questions. Key product of BI is insights and awareness of the environment with adequate coverage. For their production BI uses advanced methods and techniques—analytical, modeling and other advanced software; a variety of data collections and information sources; communication and presentation techniques; also advanced managerial approaches. In retrospect of research efforts, the technology part received more attention and is more researched than the managerial and human part of BI activities. This should explain many cases of BI failure and disappointment. However, lately the research interest for human factors in BI appears to be on the rise. **As in non-business areas of advanced informing—e.g., scientific research, military and political intelligence, users expect correct and trusted representation of issues of interest, together with adequate coverage of their context. Concentration on technical**

issues alone makes this coverage problematic, and should be balanced by additional attention to human and managerial issues.

As presented in Chap. 3, BI works in tandem with all information systems, unites their input and develops advanced results that other systems are not capable of. Systems like ERP provide initial intelligence at the levels of operations and middle-management, and their data collections and results are used further up in the insight development value chain. It has to be noted that important discoveries can be made at any level of informing activities, but it is the function of BI to integrate such discoveries into a bigger picture. Apart from internal systems, a multitude of external information sources are used and often combined with internal data to have an encompassing view of the business environment. The multiple efforts to have a complete view show the holistic nature of BI. It is mostly agreed that BI evolved from decision support, and currently encompasses decision support activities. This point is supported by one of the long-time authorities in the field of decision support, Clyde Holsapple (Holsapple, Lee-Post, & Pakath, 2014), stating that, apart from decision making, other kinds of BI activities exist—e.g., sense-making, evaluation, and prediction. There are other fields of activity that deal with intelligence, research and insights, and have accumulated experience over time. Such fields are political or military intelligence, scientific research or criminalistics—their conditions and goals differ from business intelligence, but there are multiple overlay areas that bring beneficial experience to BI. A good example is the subfield of competitive intelligence that often employs people with former experience in political or military intelligence—their experience in sensemaking from multiple sources of heterogeneous information and development of insights is very valuable in understanding competition and competitive climate. **BI may be seen as a derivative of the family of information systems in an organization, using their resources and external resources to build insights of required coverage by extracting, adding and integrating relevant information.**

Businesses have a set of dimensions that define their specific features like large or small, global or local, specialized or diverse, successful or struggling. The adequate coverage of business dimensions against limited resources of human competencies, time and money is one of the primary tasks of BI. As well, the variety of BI implementations has a number of important BI dimensions that require decisions for BI deployment: centralized or decentralized, internal or external, wide or narrow coverage, and others. This variety of BI dimensions, discussed in Chap. 4, creates numerous options for BI deployment, and some of them will better fit the needs of given business than others. An overarching controversy, featured by many systems, is also present in BI—the controversy between flexibility and effectiveness. Moore (2011) states that to be successful against strategic surprise, organizations have to be both imaginative and systematic. If intelligence is to rise above the noise and get the attention of policy, and then be acted upon, it must be both—a balance of analytical freedom and organized resources should be sought.

The available research in the relation of business dimensions and BI dimensions is rather scattered, and would benefit from more pointed research efforts. This would especially help to decide in the cases of some dimensions that are confusing, like real time BI versus right time BI; or system push versus user pull. The potentially better match of user needs and deployed BI system would also shelter users from hype around advanced BI technologies and groundless expectations. **The required encompassing coverage by BI calls for better understanding of business dimensions for a given business entity, and adequate support by available dimensions of BI deployment.**

Data integration and information integration approaches and methods, presented in Chap. 5, have to produce composite meanings. BI relations with other systems and information sources are intended to join content, starting from the level of atomic data, and going all the way to information integration—sense integration and production of composite sense, or insights. Data integration issues on the level of data structures are much more researched than information integration structures, and many research sources, announcing they have researched information integration, actually present results regarding data integration. Meanwhile, information integration, which joins meanings and context, is heuristic, non-linear, often random and vague, and significantly less researched. **Information integration is expected to provide the required completeness in informing, and, due to its vagueness, has to balance formal methods providing hard integration with heuristic methods providing soft integration.**

BI lessons and experience should provide valuable assistance in cases requiring previous experience—this set of issues is discussed in Chap. 6. On one hand, the need for recorded experience and lessons learned is evident, and sometimes vital. On the other hand, there is a rather mixed experience from the development and implementation of knowledge management systems (actually—advanced information management systems), who have many common features with collections of experience and lessons learned. However advanced the platforms for knowledge management might have been, lack of motivation and unclear benefits discouraged many would-be customers from this type of systems. A controversy emerges quite often between an acute need for specific earlier experience in difficult situations, and reluctance to record and preserve this experience. Keeping this controversy and failures in mind, LL systems should not create an additional burden in information management activities—the LL environment should be kept straight and simple. Their coverage should be adequate to cover important experience with adequate features to define experience context. Apparently, more research is required to define the issues of adequate content and context for LL applications. **Regardless of contributor reluctance, collections of lessons learned add value by providing coverage of important past situations and their outcomes as a source of context and objectivity for current situations.**

BI has always been a testbed for all advanced informing technologies working towards insight development. The available array of BI technologies, briefly discussed in Chap. 7, has to provide necessary coverage for chosen BI dimensions. BI will use any of the advanced technologies existing at the moment, limited only by resources available. Together with all sophisticated technologies, a significant deal of BI is performed using rather simple and ubiquitous technologies like MS Excel or other self-service tools, originally intended for personal productivity. However vast the array of information technologies may be, it contains no "magic wand" that will solve all or majority of BI tasks. This understanding might be rather important for practitioners when trying to make decisions regarding new technologies and their potential value. It is probably rather safe to state that the influx of new types of information technologies will not slow down, considering the relative ease of their creation and innovations. In this case, any principles or methods that assist understanding the value of new technologies, would be very welcome. **The variety of BI technology instruments, regardless of their complexity or specificity, is intended to cover the entire variety of intelligence information needs.**

The discussion on BI maturity in Chap. 8 prompts an idea that the final stage of BI maturity is very hard to define or does not exist at all; and the whole idea of BI maturity is more intended to reflect the stages of acceptance. Maturity is not a technical term—an organization, being BI-mature in technical terms, will adopt BI innovations, while BI itself as an informing environment will never be mature. Maturity models may be seen as a series of attempts to have a better understanding of BI adoption and development dynamics. Following this logic, the ultimate stage of BI maturity would be not so much a perfect "maturity" stage, as a stage showing readiness to address inevitable changes and transform itself. The notion of BI agility, discussed in the same chapter, defines exactly that—being able to adapt, exploit opportunities and maintain key competencies, thus supporting organizational agility. An agile BI system avoids becoming obsolete by leveraging BI features that are vital to activities of advanced informing, yet indifferent to the changing business and technology climate. **The notions of maturity and agility encompass the stages of BI development in an organization, as well as changes and dynamics regarding BI buildup and adaptivity. While maturity models extrapolate the current experience and understanding of BI dynamics, BI agility utilizes key BI strengths that should stand the test of time.**

BI culture, discussed in Chap. 9, is seen as an environment of conventions that is part of information culture, the latter being the part of organizational culture. BI culture has to facilitate advanced informing by removing limiting managerial and other barriers. The presented notion of BI culture rests on the community of people involved in BI activities, communication and sharing of information and insights, preservation of accumulated experience, and the ability of the community to be self-supportive in terms of its growth and professional strength. It also should remove fear of making errors, and fear of disclosure of inconvenient information—the BI culture should encourage disclosure as beneficial. It might appear risky to have a powerful organization within an organization, like some sort of alternative well-informed structure. However, the issues of intelligence community, and its co-existence with official structures, are researched in other areas of intelligence activities (see, for example, Johnston (2005)). **BI culture facilitates collaboration, community and information sharing; these traits in their own turn lay important foundations for encompassing information management and informing activities.**

The usual BI resources—people, structures, rules, content, technology—are well-known and relatively well researched. In addition to them, several soft factors or assets like attention, sense, or trust, are more and more recognized as being important in any aspect of BI activities. The discussion of these factors in Chap. 10 points out to their measurement and management problems. Although the importance of the above soft factors is evident, their features and relations are rather under-researched. The complex information space, further aggravated by complicated technology landscape, raises serious challenges for protecting and developing the above-mentioned soft factors. Their prevention from erosion requires additional efforts from BI architects, developers, consultants and professional users. **The perceived importance of soft factors like attention, sense and trust should make BI coverage more complete by separating important signals from noise, navigating sensemaking and preventing confusion, testing and reinforcing trust.**

The current chapter rounds up the ideas presented in the book, and also considers the issues that weren't given attention in the previous chapters—BI education and competency development, including importance of soft competencies, and research directions coming out of interplay between book chapters.

11.3   Important Issues in BI Education

The understanding of human factors in BI activities extends to people who develop BI technologies or deal with their implementation and adoption. Their competencies, as numerous research sources presented below have shown, experience the same imbalance between technical and human factors. To master human factors better, the education process for BI professionals should accentuate more the development of soft competencies for IS and BI education programs. In an information systems project, soft skills matter the most on the business part of the project, where rich meanings are exchanged, and this exchange is a fruitful soil for misunderstandings. The business side covers ideas, concepts, expectations, communication and other activities carrying rich sense. Jung and Lehrer (2017) stress social and personal skills as a facilitator of translation between business and technical/engineering languages. Technical skills are mostly required at the engineering part of the project. For BI specialists, an important skill would be the ability to communicate the complicated benefits of BI.

The lack of soft skills is supported in (Beard, Schwieger, & Surendran, 2008): while the ISA Certification exam is oriented to the technical skill set of information systems majors, authors reflect the concerns of employers regarding the apparent insufficient competency in soft skills. The importance of managerial competencies is pointed out in the description of Maryville university undergraduate program for systems analysts (How to Become a Systems Analyst, 2020) that stresses ethics, motivation, team building and leadership. The description mentions that certification programs certainly require high level of technical knowledge, but at the same time becoming a mature systems analyst or BI professional requires adequate competency in such soft skills as effective communication and creativity.

The fact that certification programs lean towards technical skills can be explained by hard skills being universal and easy to test, and soft skills being vague and much more hard to measure. This may also reflect the expectation that soft skills will be best developed in the workplace.

Soft competences, or lack of them, have initiated discussions on educational and training issues for their development. Woodward, Sendall, and Ceccucci (2010) stated that university programs tend to lag behind industry, and presented an instructional module to promote soft skills by performing a team project in an environment that is close to IS workplace. Such project-based learning has for a long time been recognized as rich in authentic learning experiences while performing complex tasks. The similar point of giving high priority to real-life project tasks and class discussions in in class delivery of soft skills is supported by Jung and Lehrer (2017), who also present soft skills for IT managers, business consultants, technology entrepreneurs and process managers. While addressing the competencies for software developers, Faheem, Capretz, Bouktif, and Campbell (2013) accentuated the importance of evaluating the complexity of human personality leading to intricate dynamics that "cannot be ignored but which have been often overlooked". Carvalho, Sousa, and Sá (2010) have researched actual responses of students in the Information Systems Development course that were working on course team projects, where such soft issues as leadership, teamwork, presentation and communication are deemed very important.

To distinguish between BI generalists (business-oriented) and BI specialists (technology-oriented), Wixom et al. (2014) have presented a typology of BI-savvy professionals (Table 11.1):

From the above table, we can see that, although the separation of business and IT competencies is quite clear, soft skills are clearly under-represented. In addition to this, a latent emphasis is made on demand of technical skills, assuming that BI generalists who understand business are abundant, while technical BI specialists are few, and therefore in greater demand.

Once again, regarding education, BI has to turn for experience to intelligence activities in other fields. IAFIE (International Association for Intelligence Education) publishes standards for intelligence education on both graduate and undergraduate levels, where soft competencies dominate the list of degree outcomes (Intelligence education standards (n.d.)):

1. Employ knowledge of mathematics and science.
2. Identify, describe and critically evaluate applicable intelligence technologies.
3. Demonstrate the ability to professionally speak, read and orally comprehend a foreign language—as applicable.
4. Identify professional ethics, and how they apply to the intelligence field.
5. Develop general professional written and oral reports and presentations.
6. Demonstrate the ability to work collaboratively in diverse groups.
7. Demonstrate intelligence knowledge, skills and abilities in a non-academic

Table 11.1 A Typology of BI-Savvy Professionals (Wixom et al., 2014)

	BI generalist	BI specialist
Focus	Apply general BI concepts and techniques, i.e., "data literacy" to functional area, e.g., marketing, logistics, finance, etc.	Generate creative BI solutions across disciplines, i.e., "data expertise"
Typical background	Functional, e.g., marketing, logistics, finance, etc.	Technical, e.g., information systems, operations research, computer science, statistics, applied mathematics, etc.
Size of the population	Very large—typically every business major or MBA student, as well as non-business major in professional fields such as medicine, design, and production	Small—needs to grow
Emphasis on BI skills	Problem domain (business) and analytical (quantitative)	Analytical and technical
Current offerings	BI course, BI components in a generic IS or operations class, or BI minor	BI concentration or dedicated BI program
Future prospects	Employment by traditional employers, yet better value proposition from graduates	Employment in large organizations (such as the federal government() and Fortune 1000 companies, consulting firms, software developers, and academia
Next steps for academia	Make BI an integral part of every business curriculum, ideally through a dedicated course that covers the basics of data warehousing, data mining, basic statistics, and business strategy. Target business and non-business majors in traditional academic programs, as well as professional and certificate programs	Develop more BI concentrations (minors and majors) with emphasis on quantitative skills, (big) data management proficiency, systems integration, and broad understanding of organizations and their strategy. Emphasize nontraditional application areas such as healthcare management, sports management, and urban planning

setting through an internship, cooperative or supervised experience.

8. Evaluate intelligence issues or challenges through either a capstone practicum or undergraduate thesis.
9. Appraise contemporary and emergent threats, challenges and issues to business, law enforcement, homeland security, national security and regional studies spheres—as applicable.
10. Explain application of intelligence strategies and operations to business, law enforcement, homeland security, national security and regional studies issues— as applicable.

A clear emphasis on insight development dominates the above list, while in the list of current data science certification programs (White 2019), very few program descriptions mention anything about insight development skills, or at least skills to ask good business questions to data.

11.4   BI Research Issues

An encompassing nature of BI, to author's opinion, can be best achieved by symbiotic interaction between users and system, or people and technology—hybrid intelligence, augmented intelligence or whatever name can be assigned to it. Without proper attention to this interaction, BI still is and probably will be managed as an IT initiative or a regular information system (Williams, 2016), in accordance with policies, resource strategies, optimization of capacity and control. In such case, BI potential to facilitate the best use of company data risks to be significantly wasted. A reference point here may be the routine or non-routine nature of informing activities (Gill, 2015). Compared to regular information systems, whose activities are mostly routine, BI activities, while carrying their own share of routine, are expected to support informing and problem solving in non-routine situations. While classical decision support applications were meant exactly to provide support in non-routine situations, the current informing climate that merges environment monitoring and decision support needs attention from researchers to develop approaches for balancing routine and non-routine informing.

An important and potentially fruitful area of BI research is BI culture, of which fragmented discussions and mentions can be found in the BI research literature of the last decade, sometimes in the context of organizational and information culture, and sometimes as a research object of its own. These fragments, as well as author's own research, presented in Chap. 9 of this book, justify the distinction of business intelligence culture as a separate concept that unifies the scattered definitions of cultural issues relating to business intelligence in published research.

BI systems have to improve their agility to withstand permanent changes in business environment, such as new business models, technologies, data and information sources, and global shakeups like COVID-19 pandemic. This agility should ensure flexible upgrades of BI environment in the face of changes, and preservation of the key competencies. Here, the dependence of BI agility upon BI culture and organizational culture is suggested by many research sources; yet research of these dependencies is scarce. It is obvious that a rigid or complacent organizational culture limits organizational agility and BI agility at the same time. On the other hand, agile BI may be an important precondition for an agile organization, and dependencies of this type deserve research of their own.

BI technology is constantly evolving and being perfected, but it is always followed by a cloud of hype which focuses on technology's virtues, without much considering human and managerial issues regarding use of that technology. Businesses have to sort out through the deluge of technology information to separate hype from innovations that would be beneficial to their context. Research concentrating on better estimation of organization needs in BI area, and potential value of future BI systems, would be helpful for practitioners in coping with hype. There's a plethora of professional and semi-professional sources providing all sorts of advice in BI technology decisions, and there's significantly less empirical academic work in the area of BI value estimation.

A logical direction of research work in BI area would be a development of a framework like business analytics capability framework, presented by Cosic, Shanks, and Maynard (2012). However, the presented models are oriented to element maturity (governance, culture, technology, people) and optimization, which is a risky approach because of temporary nature of optimization. It is interesting to note that the next paper by same authors on the same topic (Cosic, Shanks, & Maynard, 2015) has substituted "maturity" for "capability", BACMM (business analytics capability maturity model) for BACF (business analytics capability framework), and accentuates dynamic capabilities as the base concept for their model. To author's opinion, the environment of BI is much more complicated than a model of one-way constructs, presented in many empirical works. The traditional way of empiric research (construct—hypothesis—testing), although possessing the required rigor, may be insufficient to capture deeper dependencies between such amorphous issues as agility, culture, attention, sense or trust.

The information systems research community, including the researchers working in BI area, has for a long time experienced the difficulty of balancing research **relevance** and **rigor** (O'Donnell, Sipsma, & Watt, 2012). Relevance here is seen as ideas and findings that push theory boundaries, have value for practitioners and generally impact the field of informing. Academic rigor, on the other hand, secures serious academic recognition and publication in well regarded peer-reviewed journals. The same source points out that finding the right balance between rigor and relevance is a difficult topic and has been debated extensively for a prolonged time; a special issue on this debate has been published by MIS Quarterly in 1999. There was, and still is, an impression that researchers in the field of IS and its subfields, BI notwithstanding, produce research that leans much more on rigor than relevance. While the renowned decision support systems researcher Peter Keen (1991) argued that the rigor of IS research was only important if the work was also relevant, within the sub-field of decision support systems researchers have not placed a high priority on conducting research with a focus on relevance. Keen worried in 1991 that the field was 'in danger of talking mainly to itself about itself' (O'Donnell et al., 2012). Arnott and Pervan (2014), in their recent wide-ranging empirical analysis of the literature in the area, have also noted that the majority of academic research into decision support systems isn't directly relevant to BI practitioners, and urged researchers to focus more on relevance. This point has also been supported by Gill (2015).

One of the key research issues, pertinent not only to BI field, might be the selection of a good research question to begin with. Papers with minor or vague research questions and perfect rigor should be criticized the same way as papers with ambitious research questions and liberal use of research rigor. It is the hope of this author that the issues raised in this book will merit research efforts that will be both relevant and rigorous.

References

Arnott, D., & Pervan, G. (2014). A critical analysis of decision support systems research revisited: The rise of design science. *Journal of Information Technology, 29,* 269–293.

Beard, D., Schwieger, D., & Surendran, K. (2008). Integrating soft skills assessment through university, college, and programmatic efforts at an AACSB accredited institution. *Journal of Information Systems Education, 19*(2), 229–240.

Carvalho, J.A., Sousa, R.D., & Sá, J.O. (2010). Information systems development course: Integrating business, IT and IS competencies. *Transforming Engineering Education: Creating Interdisciplinary Skills for Complex Global Environments,* Dublin, Ireland, 6–9 April, 2010.

Cosic, R., Shanks, G., & Maynard, S. (2012). Towards a business analytics capability maturity model. *Proceedings of the 23rd Australasian Conference on Information Systems.*

Cosic, R., Shanks, G., & Maynard, S. (2015). A business analytics capability framework. *Australasian Journal of Information Systems, 19,* S5–S19.

Faheem, A., Capretz, L. F., Bouktif, S., & Campbell, P. (2013). Soft skills and software development: A reflection from software industry. *International Journal of Information Processing and Management, 4*(3), 171–191.

Gill, T. G. (2015). *Informing science. Concepts and systems. Vol. 1.* Santa Rosa, CA: Informing Science Press.

Holsapple, C., Lee-Post, A., & Pakath, R. (2014). A unified foundation for business analytics. *Decision Support Systems, 64,* 130–141.

How to Become a Systems Analyst. (2020). *Maryville University, MO, USA.* Retrieved June 20, 2020, from https://online.maryville.edu/online-bachelors-degrees/management-information-systems/careers/become-a-systems-analyst/.

Intelligence Education Standards. (n.d.). *International Association for Intelligence Education.* Retrieved June 20, 2020, from www.Iafie.org/page/IntelEd.

Johnston, R. (2005). *Analytic culture in the US intelligence community.* Washington, DC: Center for the Study of Intelligence.

Jung, R., & Lehrer, C. (2017). Guidelines for education in business and information systems engineering at tertiary institutions. *Business Information Systems Engineering, 59*(3), 189–203.

Keen, P. (1991). *Shaping the future: Business design through information technology.* Cambridge, MA: Harvard University Press.

March, S., & Hevner, A. (2007). Integrated decision support systems: A data warehousing perspective. *Decision Support Systems, 43,* 1031–1043.

Moore, D. T. (2011). *Sensemaking. A structure for intelligence revolution.* Washington, DC: National Defense Intelligence College.

O'Donnell, P., Sipsma, S., & Watt, C. (2012). The critical issues facing business intelligence practitioners. *Journal of Decision Systems, 21*(3), 203–216.

Parr Rud, O. (2009). *Business intelligence success factors: Tools for aligning your business in the global economy.* Hoboken, NJ: John Wiley & Sons.

White, S. (2019). *15 Data science certifications that will pay off. CIO, December 5, 2019.* Retrieved June 20, 2020, from https://www.cio.com/article/3222879/15-data-science-certifications-that-will-pay-off.html.

Williams, S. (2016). *Business intelligence strategy and Big Data analytics. A general management perspective.* Cambridge, MA: Morgan Kaufmann.

Wixom, B., Ariyachandra, T., Douglas, D., Goul, M., Gupta, B., Iyer, L., Kulkarni, U., Mooney, J. G., Phillips-Wren, G., & Turetken, O. (2014). The current state of business intelligence in academia: The arrival of Big Data. *Communications of the Association for Information Systems, 34,* 1.

Woodward, B., Sendall, P., & Ceccucci, W. (2010). Integrating soft skill competencies through project-based learning across the information systems curriculum. *Information Systems Education Journal, 8*(8), 3–15.

Printed in the United States
by Baker & Taylor Publisher Services